Fundamentals of Aerodynamics

ndamentals of Aerodynamics
lited by Solomon Hershel
BN: 978-1-63549-687-1 (Hardback)

Larsen & Keller

blished by Larsen and Keller Education,
Penn Plaza,
th Floor,
ew York, NY 10001, USA

Cataloging-in-Publication Data

Fundamentals of aerodynamics / edited by Solomon Hershel.
p. cm.
Includes bibliographical references and index.
ISBN 978-1-63549-687-1
1. Aerodynamics. 2. Dynamics. I. Hershel, Solomon.
A358 .F86 2018
29.132 3--dc23

rinted and bound in China.

or more information regarding Larsen and Keller Education and its products, please visit the publisher's vebsite www.larsen-keller.com

Fundamentals
Aerodynam

Ed
Solomon I

Larsen & Keller
www.larsen-keller.com

Table of Contents

Preface

Aerodynamics is a sub-field of gas dynamics and fluid dynamics which deals with the study of air, particularly its relationship with any solid and fluid object. It is mainly used in the construction, design and building of airplanes, space jets, rockets and other types of flying objects. It also plays a critical role in automotive designing. This book presents the complex subject of aerodynamics in the most comprehensible and easy to understand language. It is a valuable compilation of topics, ranging from the basic to the most complex theories and principles in this field. This textbook will serve as a valuable source of reference for those interested in this field.

To facilitate a deeper understanding of the contents of this book a short introduction of every chapter is written below:

Chapter 1- Wind tunnels are used to test and scale models of aircrafts and spacecrafts. Powerful fans are installed within wind tunnels in order to move air through the tube. Improved understanding of airplanes and better air transportation are some of the reasons why wind tunnels are used. This chapter will provide an integrated understanding of wind tunnels.

Chapter 2- The instrument which is used to replicate an explosion in order to understand its effects is known as a shock tube. It is usually rectangular in shape and is made up of metal. Shock tubes are used to create and direct blast waves towards a particular object to understand the intensity of the explosion. The major components of shock tube are discussed in this section.

Chapter 3- The main instrument used in measuring pressure is called a vacuum gauge. It is used to measure pressure which is lower than the surrounding atmospheric pressure. Other scientific instruments explained in the following chapter are pressure sensor, McLeod gauge, Pirani gauge, hot-filament ionization gauge and thermocouple gauge. The topics discussed in the chapter are of great importance to broaden the existing knowledge on aerodynamics.

Chapter 4- Temperature can be measured by several scales and units. One of the devices used for temperature sensors is thermistor. Laws of thermocouples are an important part of the following section. The aspects elucidated in this section are of vital importance, and provide a better understanding of temperature measurement.

Chapter 5- The instrument used to measure velocity is known as a pitot tube. It is used to measure the speed of an aircraft, boat and the flow of gas in industrial applications. Anemometer, laser Doppler velocimetry and relativistic Doppler effect are some of the topics discussed in the following chapter. The aspects elucidated in this chapter are of vital importance, and provide a better understanding of aerodynamics.

Finally, I would like to thank the entire team involved in the inception of this book for their valuable time and contribution. This book would not have been possible without their efforts. I would also like to thank my friends and family for their constant support.

Editor

1

Understanding Wind Tunnels

Wind tunnels are used to test and scale models of aircrafts and spacecrafts. Powerful fans are installed within wind tunnels in order to move air through the tube. Improved understanding of airplanes and better air transportation are some of the reasons why wind tunnels are used. This chapter will provide an integrated understanding of wind tunnels.

Wind Tunnel

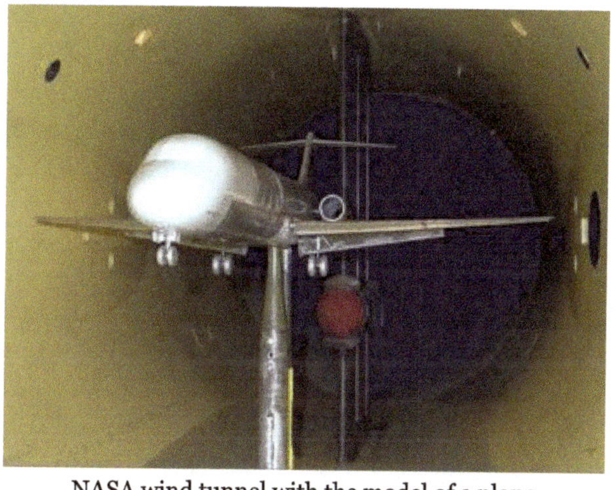

NASA wind tunnel with the model of a plane

A wind tunnel is a tool used in aerodynamic research to study the effects of air moving past solid objects. A wind tunnel consists of a tubular passage with the object under test mounted in the middle. Air is made to move past the object by a powerful fan system or other means. The test object, often called a wind tunnel model, is instrumented with suitable sensors to measure aerodynamic forces, pressure distribution, or other aerodynamic-related characteristics.

The earliest wind tunnels were invented towards the end of the 19th century, in the early days of aeronautic research, when many attempted to develop successful heavier-than-air flying machines. The wind tunnel was envisioned as a means of reversing the usual paradigm: instead of the air standing still and an object moving at speed through it, the same effect would be obtained if the object stood still and the air moved at speed past it. In that way a stationary observer could study the flying object in action, and could measure the aerodynamic forces being imposed on it.

The development of wind tunnels accompanied the development of the airplane. Large wind tunnels were built during World War II. Wind tunnel testing was considered of strategic importance during the Cold War development of supersonic aircraft and missiles.

Later on, wind tunnel study came into its own: the effects of wind on man made structures or objects needed to be studied when buildings became tall enough to present large surfaces to the wind, and the resulting forces had to be resisted by the building's internal structure. Determining such forces was required before building codes could specify the required strength of such buildings and such tests continue to be used for large or unusual buildings.

A model Cessna with helium-filled bubbles showing pathlines of the wingtip vortices

Still later, wind-tunnel testing was applied to automobiles, not so much to determine aerodynamic forces *per se* but more to determine ways to reduce the power required to move the vehicle on roadways at a given speed. In these studies, the interaction between the road and the vehicle plays a significant role, and this interaction must be taken into consideration when interpreting the test results. In an actual situation the roadway is moving relative to the vehicle but the air is stationary relative to the roadway, but in the wind tunnel the air is moving relative to the roadway, while the roadway is stationary relative to the test vehicle. Some automotive-test wind tunnels have incorporated moving belts under the test vehicle in an effort to approximate the actual condition, and very similar devices are used in wind tunnel testing of aircraft take-off and landing configurations.

The advances in computational fluid dynamics (CFD) modelling on high speed digital computers has reduced the demand for wind tunnel testing. However, CFD results are still not completely reliable and wind tunnels are used to verify CFD predictions.

Measurement of Aerodynamic Forces

Air velocity and pressures are measured in several ways in wind tunnels.

Air velocity through the test section is determined by Bernoulli's principle. Measurement of the dynamic pressure, the static pressure, and (for compressible flow only) the temperature rise in the airflow. The direction of airflow around a model can be determined by tufts of yarn attached to the aerodynamic surfaces. The direction of airflow approaching a surface can be visualized by mounting threads in the airflow ahead of and aft of the test model. Smoke or bubbles of liquid can be introduced into the airflow upstream of the test model, and their path around the model can be photographed.

Aerodynamic forces on the test model are usually measured with beam balances, connected to the test model with beams, strings, or cables.

The pressure distributions across the test model have historically been measured by drilling many small holes along the airflow path, and using multi-tube manometers to measure the pressure at each hole. Pressure distributions can more conveniently be measured by the use of pressure-sensitive paint, in which higher local pressure is indicated by lowered fluorescence of the paint at that point. Pressure distributions can also be conveniently measured by the use of pressure-sensitive pressure belts, a recent development in which multiple ultra-miniaturized pressure sensor modules are integrated into a flexible strip. The strip is attached to the aerodynamic surface with tape, and it sends signals depicting the pressure distribution along its surface.

Pressure distributions on a test model can also be determined by performing a wake survey, in which either a single pitot tube is used to obtain multiple readings downstream of the test model, or a multiple-tube manometer is mounted downstream and all its readings are taken.

The aerodynamic properties of an object can not all remain the same for a scaled model. However, by observing certain similarity rules, a very satisfactory correspondence between the aerodynamic properties of a scaled model and a full-size object can be achieved. The choice of similarity parameters depends on the purpose of the test, but the most important conditions to satisfy are usually:

- Geometric similarity: all dimensions of the object must be proportionally scaled;
- Mach number: the ratio of the airspeed to the speed of sound should be identical for the scaled model and the actual object (having identical Mach number in a wind tunnel and around the actual object is -not- equal to having identical airspeeds);
- Reynolds number: the ratio of inertial forces to viscous forces should be kept. This parameter is difficult to satisfy with a scaled model and has led to development of pressurized and cryogenic wind tunnels in which the viscosity of the working fluid can be greatly changed to compensate for the reduced scale of the model.

In certain particular test cases, other similarity parameters must be satisfied, such as e.g. Froude number.

History

English military engineer and mathematician Benjamin Robins (1707–1751) invented a whirling arm apparatus to determine drag and did some of the first experiments in aviation theory.

Sir George Cayley (1773–1857) also used a whirling arm to measure the drag and lift of various airfoils. His whirling arm was 5 feet (1.5 m) long and attained top speeds between 10 and 20 feet per second (3 to 6 m/s).

However, the whirling arm does not produce a reliable flow of air impacting the test shape at a normal incidence. Centrifugal forces and the fact that the object is moving in its own wake mean that detailed examination of the airflow is difficult. Francis Herbert Wenham (1824–1908), a Council Member of the Aeronautical Society of Great Britain, addressed these issues by inventing, designing and operating the first enclosed wind tunnel in 1871. Once this breakthrough had been achieved, detailed technical data was rapidly extracted by the use of this tool. Wenham and his colleague John Browning are credited with many fundamental discoveries, including the measurement of l/d ratios, and the revelation of the beneficial effects of a high aspect ratio.

Konstantin Tsiolkovsky built an open-section wind tunnel with a centrifugal blower in 1897, and determined the drag coefficients of flat plates, cylinders and spheres.

Danish inventor Poul la Cour applied wind tunnels in his process of developing and refining the technology of wind turbines in the early 1890s.

Carl Rickard Nyberg used a wind tunnel when designing his *Flugan* from 1897 and onwards.

In a classic set of experiments, the Englishman Osborne Reynolds (1842–1912) of the University of Manchester demonstrated that the airflow pattern over a scale model would be the same for the full-scale vehicle if a certain flow parameter were the same in both cases. This factor, now known as the Reynolds number, is a basic parameter in the description of all fluid-flow situations, including the shapes of flow patterns, the ease of heat transfer, and the onset of turbulence. This comprises the central scientific justification for the use of models in wind tunnels to simulate real-life phenomena. However, there are limitations on conditions in which dynamic similarity is based upon the Reynolds number alone.

Replica of the Wright brothers' wind tunnel

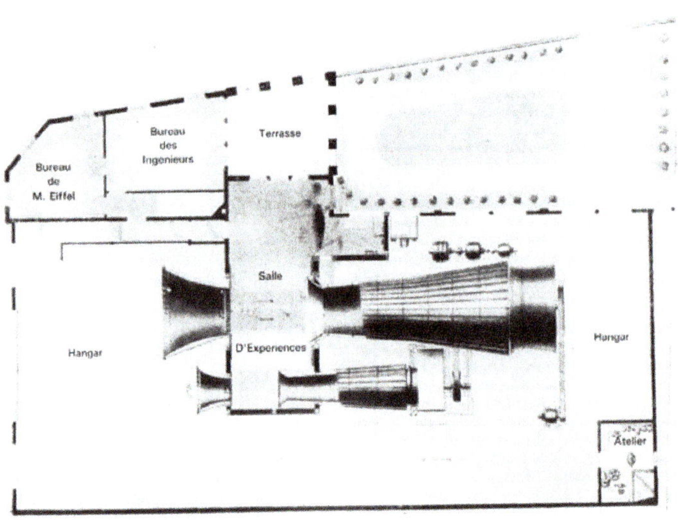

Eiffel's wind tunnels in the Auteuil laboratory

German aviation laboratory, 1935

The Wright brothers' use of a simple wind tunnel in 1901 to study the effects of airflow over various shapes while developing their Wright Flyer was in some ways revolutionary. It can be seen from the above, however, that they were simply using the accepted technology of the day, though this was not yet a common technology in America.

In France, Gustave Eiffel (1832-1923) built his first open-return wind tunnel in 1909, powered by a 50 kW electric motor, at Champs-de-Mars, near the foot of the tower that bears his name. Between 1909 and 1912 Eiffel ran about 4000 tests in his wind tunnel, and his systematic experimentation set new standards for aeronautical research. In 1912 Eiffel's laboratory was moved to Auteuil, a suburb of Paris, where his wind tunnel with a 2-metre test section is still operational today. Eiffel significantly improved the efficiency of the open-return wind tunnel by enclosing the test section in a chamber, designing a flared inlet with a honeycomb flow straightener and adding a diffuser between the test section and the fan located at the downstream end of the diffuser; this was an arrangement followed by a number of wind tunnels later built; in fact the open-return low speed wind tunnel is often called the Eiffel-type wind tunnel.

Subsequent use of wind tunnels proliferated as the science of aerodynamics and discipline of aeronautical engineering were established and air travel and power were developed.

The US Navy in 1916 built one of the largest wind tunnels in the world at that time at the Washington Navy Yard. The inlet was almost 11 feet (3.4 m) in diameter and the discharge part was 7 feet (2.1 m) in diameter. A 500 hp electric motor drove the paddle type fan blades.

In 1931 the NACA built a 30-foot by 60-foot "full scale" wind tunnel at Langley Research Center in Langley, Virginia. The tunnel was powered by a pair of fans driven by 4000 hp electric motors. The layout was a double-return, closed-loop format and could accommodate many full-size real aircraft as well as scale models. The tunnel was eventually closed and, even though it was declared a historic national landmark in 1995, demolition began in 2010.

Until World War II, the world's largest wind tunnel, built in 1932-1934, was located in a suburb of Paris, Chalais-Meudon, France. It was designed to test full size aircraft and had six large fans driven by high powered electric motors. The Chalais Meudon wind tunnel was used by ONERA under the name S1Ch until 1976, e.g. in the development of the Caravelle and Concorde airplanes. Today, this wind tunnel is preserved as a national monument.

Ludwig Prandtl was Theodore von Kármán's teacher at Gottingen University and suggested the construction of a wind tunnel for tests of airships they were designing. The vortex street of turbulence downstream of a cylinder was tested in the tunnel. When he later moved to Aachen University he recalled use of this facility:

> I remembered the wind tunnel in Gottingen was started as a tool for studies of Zeppelin behavior, but that it had proven to be valuable for everything else from determining the direction of smoke from a ship's stack, to whether a given airplane would fly. Progress at Aachen, I felt, would be virtually impossible without a good wind tunnel.

When von Kármán began to consult with Caltech he worked with Clark Millikan and Arthur L. Klein. He objected to their design and insisted on a return flow making the device "independent of the fluctuations of the outside atmosphere". It was completed in 1930 and used for Northrop Alpha testing. In 1939 General Arnold asked what was required to advance the USAF, and von Kármán answered, "The first step is to build the right wind tunnel." On the other hand, after the successes of the Bell X-2 and prospect of more advanced research, he wrote, "I was in favor of constructing such a plane because I have never believed that you can get all the answers out of a wind tunnel."

World War II

In 1941 the US constructed one of the largest wind tunnels at that time at Wright Field in Dayton, Ohio. This wind tunnel starts at 45 feet (14 m) and narrows to 20 feet (6.1 m) in diameter. Two 40-foot (12 m) fans were driven by a 40,000 hp electric motor. Large scale aircraft models could be tested at air speeds of 400 mph (640 km/h).

The wind tunnel used by German scientists at Peenemünde prior to and during WWII is an interesting example of the difficulties associated with extending the useful range of large wind tunnels. It used some large natural caves which were increased in size by excavation and then sealed to store large volumes of air which could then be routed through the wind tunnels. This innovative approach allowed lab research in high-speed regimes and greatly accelerated the rate of advance of Germany's aeronautical engineering efforts. By the end of the war, Germany had at least three different *supersonic* wind tunnels, with one capable of Mach 4.4 (heated) airflows.

A large wind tunnel under construction near Oetztal, Austria would have had two fans directly driven by two 50,000 horsepower hydraulic turbines. The installation was not completed by the end of the war and the dismantled equipment was shipped to Modane, France in 1946 where it was re-erected and is still operated there by the ONERA. With its 8m test section and airspeed up to Mach 1 it is the largest transonic wind tunnel facility in the world.

On June 22, 1942 Curtiss-Wright financed construction of one of the nation's largest subsonic wind tunnels in Buffalo, N.Y. The first concrete for building was poured on June 22, 1942 on a site that eventually would become Calspan, where the largest independently-owned wind tunnel in the United States still operates.

By the end of World War II, the US had built eight new wind tunnels, including the largest one in the world at Moffett Field near Sunnyvale, California, which was designed to test full size aircraft at speeds of less than 250 mph and a vertical wind tunnel at Wright Field, Ohio, where the wind stream is upwards for the testing of models in spin situations and the concepts and engineering designs for the first primitive helicopters flown in the US.

After World War II

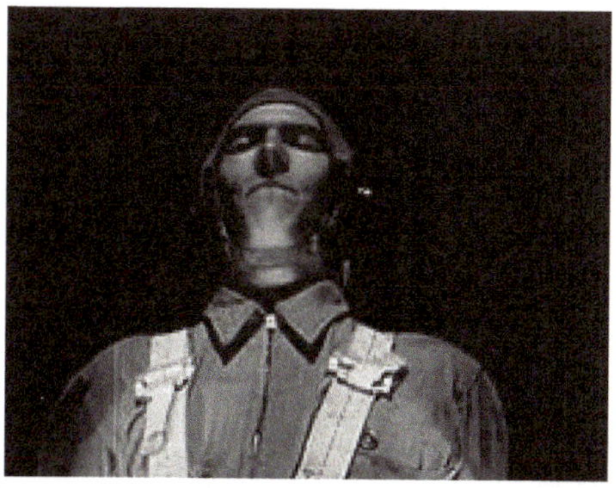

NACA wind tunnel test on a human subject, showing the effects of high wind speeds on the human face

Later research into airflows near or above the speed of sound used a related approach. Metal pressure chambers were used to store high-pressure air which was then accelerated through a nozzle designed to provide supersonic flow. The observation or instrumentation chamber ("test section") was then placed at the proper location in the throat or nozzle for the desired airspeed.

In the United States, concern over the lagging of American research facilities compared to those built by the Germans lead to the Unitary Wind Tunnel Plan Act of 1949, which authorized expenditure to construct new wind tunnels at universities and at military sites. Some German war-time wind tunnels were dismantled for shipment to the United States as part of the plan to exploit German technology developments.

Preparing a model in the Kirsten Wind Tunnel, a subsonic wind tunnel at the University of Washington

For limited applications, Computational fluid dynamics (CFD) can supplement or possibly replace the use of wind tunnels. For example, the experimental rocket plane SpaceShipOne was designed without any use of wind tunnels. However, on one test, flight threads were attached to the surface of the wings, performing a wind tunnel type of test during an actual flight in order to refine the computational model. Where external turbulent flow is present, CFD is not practical due to limitations in present-day computing resources. For example, an area that is still much too complex for the use of CFD is determining the effects of flow on and around structures, bridges, terrain, etc.

The most effective way to simulative external turbulent flow is through the use of a boundary layer wind tunnel.

There are many applications for boundary layer wind tunnel modeling. For example, understanding the impact of wind on high-rise buildings, factories, bridges, etc. can help building designers construct a structure that stands up to wind effects in the most efficient manner possible. Another significant application for boundary layer wind tunnel modeling is for understanding exhaust gas dispersion patterns for hospitals, laboratories, and other emitting sources. Other examples of boundary layer wind tunnel applications are assessments of pedestrian comfort and snow drifting. Wind tunnel modeling is accepted as a method for aiding in Green building design. For instance, the use of boundary layer wind tunnel modeling can be used as a credit for Leadership in Energy and Environmental Design (LEED) certification through the U.S. Green Building Council.

Fan blades of Langley Research Center's 16 foot transonic wind tunnel in 1990, before it was mothballed in 2004

Wind tunnel tests in a boundary layer wind tunnel allow for the natural drag of the Earth's surface to be simulated. For accuracy, it is important to simulate the mean wind speed profile and turbulence effects within the atmospheric boundary layer. Most codes and standards recognize that wind tunnel testing can produce reliable information for designers, especially when their projects are in complex terrain or on exposed sites.

In the United States, many wind tunnels have been decommissioned in the last 20 years, including some historic facilities. Pressure is brought to bear on remaining wind tunnels due to declining or erratic usage, high electricity costs, and in some cases the high value of the real estate upon which the facility sits. On the other hand, CFD validation still requires wind-tunnel data, and this is likely to be the case for the foreseeable future. Studies have been done and others are under way to assess future military and commercial wind tunnel needs, but the outcome remains uncertain. More recently an increasing use of jet-powered, instrumented unmanned vehicles ["research drones"] have replaced some of the traditional uses of wind tunnels.

How it Works

Six-element external balance below the Kirsten Wind Tunnel

Air is blown or sucked through a duct equipped with a viewing port and instrumentation where models or geometrical shapes are mounted for study. Typically the air is moved through the tunnel using a series of fans. For very large wind tunnels several meters in diameter, a single large fan is not practical, and so instead an array of multiple fans are used in parallel to provide sufficient airflow. Due to the sheer volume and speed of air movement required, the fans may be powered by stationary turbofan engines rather than electric motors.

The airflow created by the fans that is entering the tunnel is itself highly turbulent due to the fan blade motion (when the fan is blowing air into the test section – when it is sucking air out of the test section downstream, the fan-blade turbulence is not a factor), and so is not directly useful for accurate measurements. The air moving through the tunnel needs to be relatively turbulence-free and laminar. To correct this problem, closely spaced vertical and horizontal air vanes are used to smooth out the turbulent airflow before reaching the subject of the testing.

Due to the effects of viscosity, the cross-section of a wind tunnel is typically circular rather than square, because there will be greater flow constriction in the corners of a square tunnel that can make the flow turbulent. A circular tunnel provides a smoother flow.

The inside facing of the tunnel is typically as smooth as possible, to reduce surface drag and turbulence that could impact the accuracy of the testing. Even smooth walls induce some drag into the airflow, and so the object being tested is usually kept near the center of the tunnel, with an empty buffer zone between the object and the tunnel walls. There are correction factors to relate wind tunnel test results to open-air results.

The lighting is usually embedded into the circular walls of the tunnel and shines in through windows. If the light were mounted on the inside surface of the tunnel in a conventional manner, the light bulb would generate turbulence as the air blows around it. Similarly, observation is usually done through transparent portholes into the tunnel. Rather than simply being flat discs, these lighting and observation windows may be curved to match the cross-section of the tunnel and further reduce turbulence around the window.

Various techniques are used to study the actual airflow around the geometry and compare it with theoretical results, which must also take into account the Reynolds number and Mach number for the regime of operation.

Pressure Measurements

Pressure across the surfaces of the model can be measured if the model includes pressure taps. This can be useful for pressure-dominated phenomena, but this only accounts for normal forces on the body.

Force and Moment Measurements

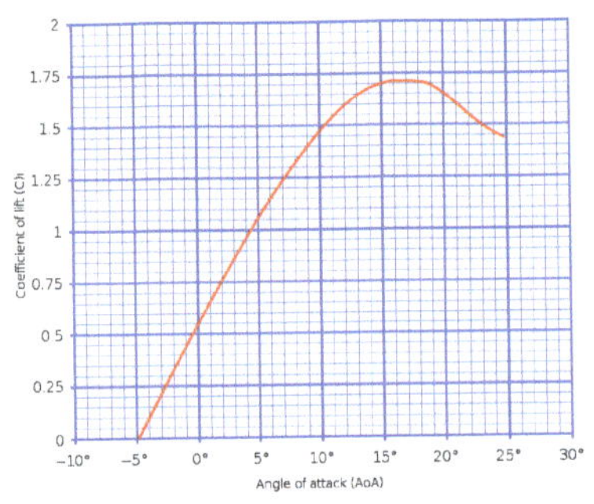

A typical lift coefficient versus angle of attack curve

With the model mounted on a force balance, one can measure lift, drag, lateral forces, yaw, roll, and pitching moments over a range of angle of attack. This allows one to produce common curves such as lift coefficient versus angle of attack.

The force balance itself creates drag and potential turbulence that will affect the model and introduce errors into the measurements. The supporting structures are therefore typically smoothly shaped to minimize turbulence.

Flow Visualization

Because air is transparent it is difficult to directly observe the air movement itself. Instead, multiple methods of both quantitative and qualitative flow visualization methods have been developed for testing in a wind tunnel.

Qualitative Methods

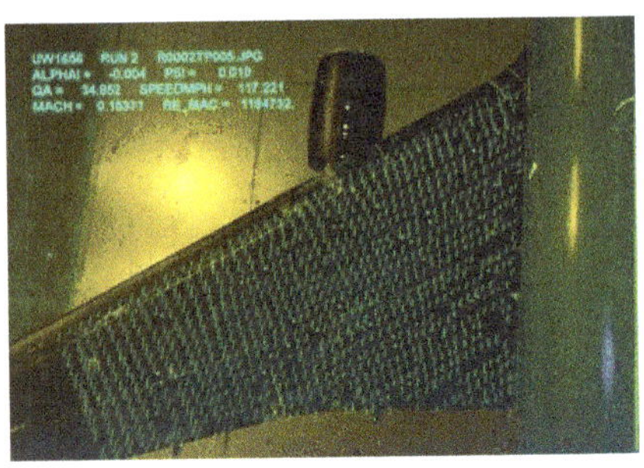

Compilation of images taken during an alpha run starting at 0 degrees alpha ranging to 26 degrees alpha. Images taken at the Kirsten Wind Tunnel using fluorescent mini-tufts. Notice how separation starts at the outboard wing and progresses inward. Notice also how there is delayed separation aft of the nacelle.

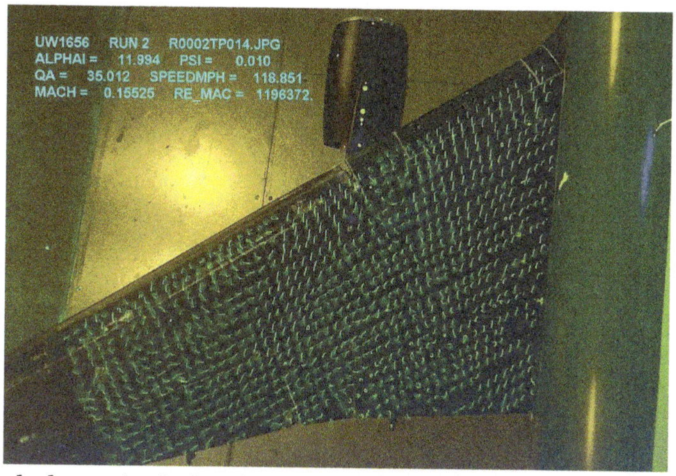

Fluorescent mini-tufts attached to a wing in the Kirsten Wind Tunnel showing air flow direction and separation. Angle of attack ~ 12 degrees, speed ~120 Mph.

China clay on a wing in the Kirsten Wind Tunnel showing reverse and span-wise flow

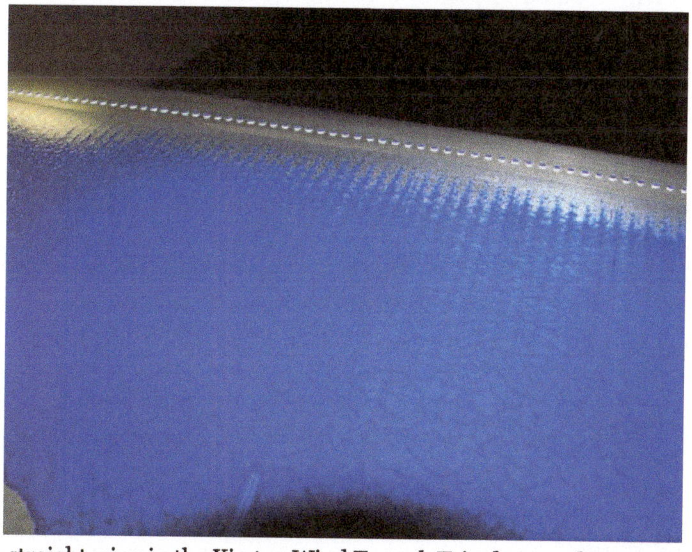

Oil flow visible on a straight wing in the Kirsten Wind Tunnel. Trip dots can be seen near the leading edge.

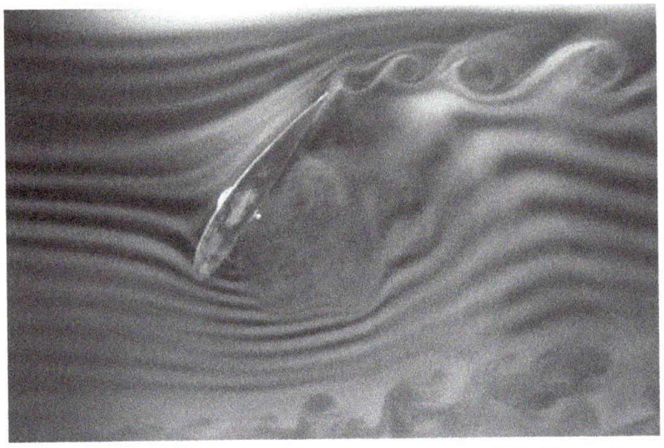

Fog (water particle) wind tunnel visualization of a NACA 4412 airfoil at a low-speed flow (Re=20.000)

- Smoke
- Tufts are applied to a model and remain attached during testing. Tufts can be used to gauge air flow patterns and flow separation.
- Evaporating suspensions are simply a mixture of some sort or fine powder, talc, or clay mixed into a liquid with a low latent heat of evaporation. When the wind is turned on the liquid quickly evaporates, leaving behind the clay in a pattern characteristic of the air flow.
- Oil: When oil is applied to the model surface it can clearly show the transition from laminar to turbulent flow as well as flow separation.
- Fog (usually from water particles) is created with an ultrasonic piezoelectric nebulizer. The fog is transported inside the wind tunnel (preferably of the closed circuit and closed test section type). An electrically heated grid is inserted before the test section, which evaporates the water particles at its vicinity, thus forming fog sheets. The fog sheets function as streamlines over the test model when illuminated by a light sheet.
- Sublimation: If the air movement in the tunnel is sufficiently non-turbulent, a particle stream released into the airflow will not break up as the air moves along, but stay together as a sharp thin line. Multiple particle streams released from a grid of many nozzles can provide a dynamic three-dimensional shape of the airflow around a body. As with the force balance, these injection pipes and nozzles need to be shaped in a manner that minimizes the introduction of turbulent airflow into the airstream.

High-speed turbulence and vortices can be difficult to see directly, but strobe lights and film cameras or high-speed digital cameras can help to capture events that are a blur to the naked eye.

High-speed cameras are also required when the subject of the test is itself moving at high speed, such as an airplane propeller. The camera can capture stop-motion images of how the blade cuts through the particulate streams and how vortices are generated along the trailing edges of the moving blade.

Classification

There are many different kinds of wind tunnels; an overview is given in the list below:

- Low-speed wind tunnel

- High speed wind tunnel
- Subsonic and transonic wind tunnel
- Supersonic wind tunnel
- Hypersonic wind tunnel

Wind tunnels are also classified based on their main use.

Aeronautical Wind Tunnels

The main subcategories in the aeronautical wind tunnels are:

High Reynolds Number Tunnels

Reynolds number is one of the governing similarity parameters for the simulation of flow in a wind tunnel. For mach number less than 0.3, it is the primary parameter that governs the flow characteristics. There are three main ways to simulate high Reynolds number, since it is not practical to obtain full scale Reynolds number by use of a full scale vehicle.

- Pressurised tunnels: Here test gases are pressurised to increase the Reynolds number.
- Heavy gas tunnels: Heavier gases like freon and R-134a are used as test gases. The transonic dynamics tunnel at NASA Langley is an example of such a tunnel.
- Cryogenic tunnels: Here test gas is cooled down to increase the Reynolds number. The European transonic wind tunnel uses this technique.
- High-altitude tunnels: These are designed to test the effects of shock waves against various aircraft shapes in near vacuum. In 1952 the University of California constructed the first two high-altitude wind tunnels: one for testing objects at 50 to 70 miles above the earth and the second for tests at 80 to 200 miles above the earth.

V/STOL Tunnels

V/STOL tunnels require large cross section area, but only small velocities. Since power varies with the cube of velocity, the power required for the operation is also less. An example of a V/STOL tunnel is the NASA Langley 14' x 22' tunnel.

Spin Tunnels

Aircraft have a tendency to go to spin when they stall. These tunnels are used to study that phenomenon.

Automotive Tunnels

Automotive wind tunnels fall into two categories:

- External flow tunnels are used to study the external flow through the chassis
- Climatic tunnels are used to evaluate the performance of door systems, braking systems,

etc. under various climatic conditions. Most of the leading automobile manufacturers have their own climatic wind tunnels

Wunibald Kamm built the first full-scale wind tunnel for motor vehicles.

For external flow tunnels various systems are used to compensate for the effect of the boundary layer on the road surface, including systems of moving belts under each wheel and the body of the car (5 or 7 belt systems) or one large belt under the entire car, or other methods of boundary layer control such as scoops or perforations to suck it away.

Aeroacoustic Tunnels

These tunnels are used in the studies of noise generated by flow and its suppression.

Vertical wind tunnel T-105 at Central Aerohydrodynamic Institute, Moscow, built in 1941 for aircraft testing

Aquadynamic Flume

The aerodynamic principles of the wind tunnel work equally on watercraft, except the water is more viscous and so sets greater forces on the object being tested. A looping flume is typically used for underwater aquadynamic testing. The interaction between two different types of fluids means that pure wind tunnel testing is only partly relevant. However, a similar sort of research is done in a towing tank.

Low-speed Oversize Liquid Testing

Air is not always the best test medium for studying small-scale aerodynamic principles, due to the speed of the air flow and airfoil movement. A study of fruit fly wings designed to understand how the wings produce lift was performed using a large tank of mineral oil and wings 100 times larger than actual size, in order to slow down the wing beats and make the vortices generated by the insect wings easier to see and understand.

Fan Testing

Wind tunnel tests are also performed to precisely measure the air movement of fans at a specific pressure. By determining the environmental circumstances during measurement, and by revising the air-tightness afterwards, the standardization of the data is ensured.

There are two possible ways of measurement: a complete fan, or an impeller on a hydraulic installation. Two measuring tubes enable measurements of lower air currents (< 30.000 m^3/h) as well as higher air currents (< 60.000 m^3/h). The determination of the Q/h curve of the fan is one of the main objectives. To determine this curve (and to define other parameters) air technical, mechanical as well as electro technical data are measured:

Air technical:

- Static pressure difference (Pa)
- Amount of moved air (m^3/h)
- Average air speed (m/s)
- Specific efficiency ($W/1000m^3/h$)
- Efficiency

Electro technical:

- Tension (V)
- Current (A)
- Cos φ
- Admitted power (W) fan / impeller
- Rotations per minute (RPM)

The measurement can take place on the fan or in the application in which the fan is used.

Wind Engineering Testing

In wind engineering, wind tunnel tests are used to measure the velocity around, and forces or pressures upon structures. Very tall buildings, buildings with unusual or complicated shapes (such as a tall building with a parabolic or a hyperbolic shape), cable suspension bridges or cable stayed bridges are analyzed in specialized atmospheric boundary layer wind tunnels. These feature a long upwind section to accurately represent the wind speed and turbulence profile acting on the structure. Wind tunnel tests provide the necessary design pressure measurements in use of the dynamic analysis and control of tall buildings.

Subsonic and Transonic Wind Tunnel

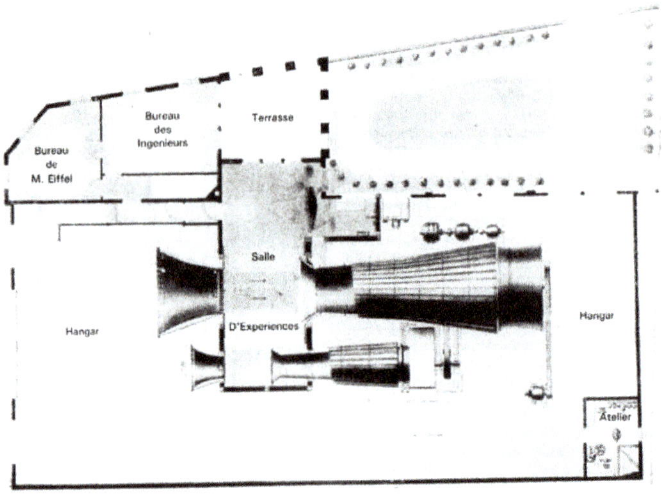

Floor plan of Eiffel's 1912 laboratory at Auteuil, Paris, with two open-return wind tunnels

Subsonic Tunnel

Low-speed wind tunnels are used for operations at very low Mach number, with speeds in the test section up to 480 km/h (~ 134 m/s, M = 0.4). They may be of open-return type (also known as the Eiffel type), or closed-return flow (also known as the Prandtl type) with air moved by a propulsion system usually consisting of large axial fans that increase the dynamic pressure to overcome the viscous losses.

Open Wind Tunnel

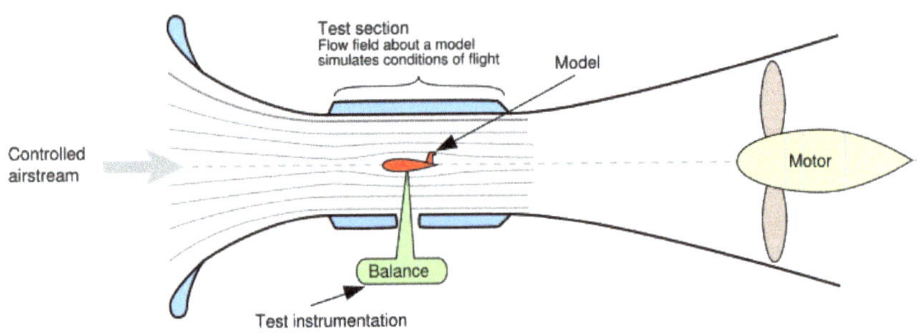

Schematic of an open wind tunnel with a closed test section

The working principle is based on the continuity and Bernoulli's equation:

The continuity equation is given by:

$$AV = constant \Rightarrow \frac{dA}{A} = -\frac{dV}{V}$$

The Bernoulli equation states:

$$P_{total} = P_{static} + P_{dynamic} = P_s + \frac{1}{2}\rho V^2$$

Putting Bernoulli into the continuity equation gives:

$$V_m^2 = 2\frac{C^2}{C^2-1}\frac{P_{settl}-p_m}{\rho} \approx 2\frac{\Delta p}{\rho}$$

The contraction ratio of a windtunnel can now be calculated by: $C = \frac{A_{settl}}{A_m}$

Closed Wind Tunnel

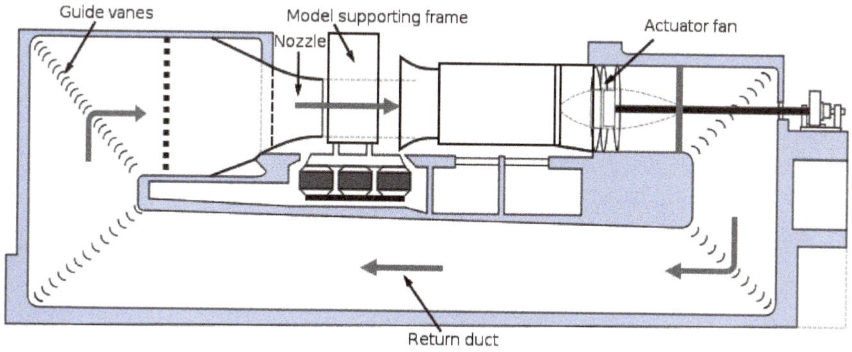

Schematic of a closed (return-flow) wind tunnel

In a return-flow wind tunnel the return duct must be properly designed to reduce the pressure losses and to ensure smooth flow in the test section. The compressible flow regime: Again with the continuity law, but now for isentropic flow gives:

$$-\frac{d\rho}{\rho} = -\frac{1}{a^2}\frac{dp}{\rho} = -\frac{1}{a^2}\frac{-\rho V dV}{\rho} = \frac{V}{a^2}dV$$

The 1-D area-velocity is known as:

$$\frac{dA}{A} = (M^2-1)\frac{dV}{V}$$

The minimal area A where M=1, also known as the *sonic throat* area is than given for a perfect gas:

$$\left(\frac{A}{A_{throat}}\right)^2 = \frac{1}{M^2}\left(\frac{2}{\gamma+1}\left(1+\frac{\gamma-1}{2}M^2\right)\right)^{\frac{\gamma+1}{\gamma-1}}$$

Transonic Tunnel

High subsonic wind tunnels (0.4 < M < 0.75) and transonic wind tunnels (0.75 < M < 1.2) are designed on the same principles as the subsonic wind tunnels. The highest speed is reached in the test section. The Mach number is approximately 1 with combined subsonic and supersonic flow regions. Testing at transonic speeds presents additional problems, mainly due to the reflection of the shock waves from the walls of the test section. Therefore, perforated or slotted walls are required to reduce shock reflection from the walls. Since important viscous or inviscid interactions occur (such as shock waves or boundary layer interaction) both Mach and Reynolds number are important and must be properly simulated. Large-scale facilities and/or pressurized or cryogenic wind tunnels are used.

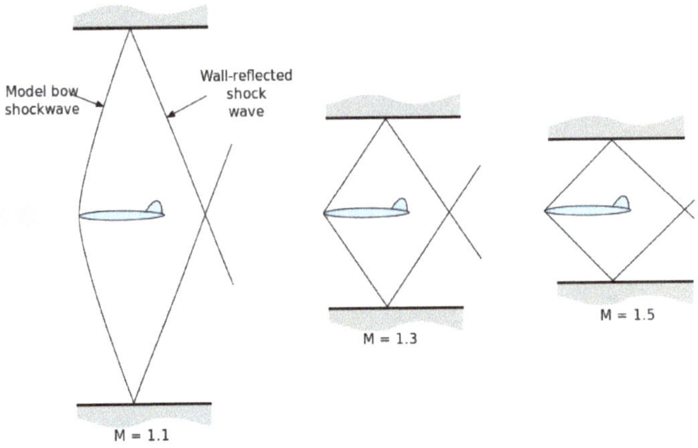

De Laval Nozzle

With a sonic throat, the flow can be accelerated or slowed down. This follows from the 1D area–velocity equation. If an acceleration to supersonic flow is required, a convergent-divergent nozzle is required. Otherwise:

- Subsonic (M < 1) then $\frac{dA}{dx} < 0 \Rightarrow$ converging
- Sonic throat (M = 1) where $\frac{dA}{dx} = 0$
- Supersonic (M >1) then $\frac{dA}{dx} > 0 \Rightarrow$ diverging

Conclusion: The Mach number is controlled by the expansion ratio $\frac{A}{A_{throat}}$.

Supersonic Wind Tunnel

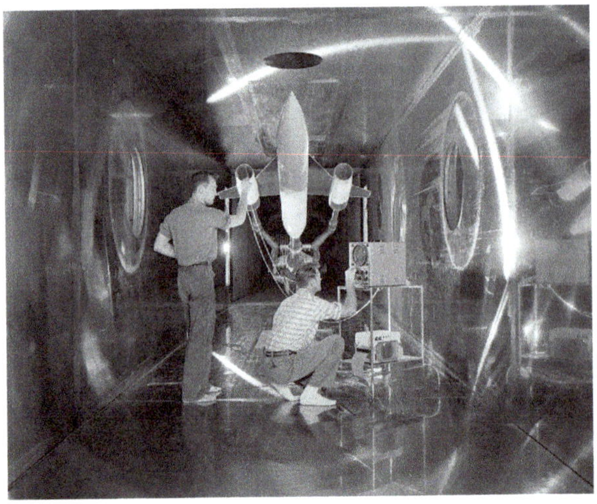

Engineers check an aircraft model before a test run in the Supersonic Wind Tunnel at Lewis Flight Propulsion Laboratory.

Schlieren photography is often used to capture images of gas flow and shock waves in supersonic wind tunnels. Here, Mach 4 flow over a pitot probe is observed by schlieren optics in the Penn State Supersonic Wind Tunnel. The flow direction is left-to-right.

A supersonic wind tunnel is a wind tunnel that produces supersonic speeds ($1.2<M<5$) The Mach number and flow are determined by the nozzle geometry. The Reynolds number is varied by changing the density level (pressure in the settling chamber). Therefore, a high pressure ratio is required (for a supersonic regime at M=4, this ratio is of the order of 10). Apart from that, condensation of moisture or even gas liquefaction can occur if the static temperature becomes cold enough. This means that a supersonic wind tunnel usually needs a drying or a pre-heating facility. A supersonic wind tunnel has a large power demand, so that most are designed for intermittent instead of continuous operation.

Restrictions for Supersonic Tunnel Operation

Minimum Required Pressure Ratio

Optimistic estimate: Pressure ratio $\leq$ the total pressure ratio over normal shock at M in test section:

$$\frac{P_t}{P_{amb}} \leq \left(\frac{P_{t_1}}{P_{t_2}}\right)_{M_1=M_m}$$

Temperature Effects: Condensation

Temperature in the test section:

$$\frac{T_m}{T_t} = \left(1+\frac{\gamma-1}{2}M_m^2\right)^{-1}$$

with $T_t = 330$ K: $T_m = 70$ K at $M_m = 4$

The velocity range is limited by reservoir temperature.

Power Requirements

The power required to run a supersonic wind tunnel is enormous, of the order of 50 MW per square meter of test section cross-sectional area. For this reason most wind tunnels operate intermittently using energy stored in high-pressure tanks. These wind tunnels are also called intermittent supersonic blowdown wind tunnels (of which a schematic preview is given below). Another way of achieving the huge power output is with the use of a vacuum storage tank. These tunnels are called indraft supersonic wind tunnels, and are seldom used because they are restricted to low Reynolds numbers. Some large countries have built major supersonic tunnels that run continuously; one is shown in the photo. Other problems operating a supersonic wind tunnel include:

- starting and unstart of the test section (related to maintaining at least a minimum pressure ratio)
- adequate supply of dry air
- wall interference effects due to shock wave reflection and (sometimes) blockage
- high-quality instruments capable of rapid measurements due to short run times in intermittent tunnels

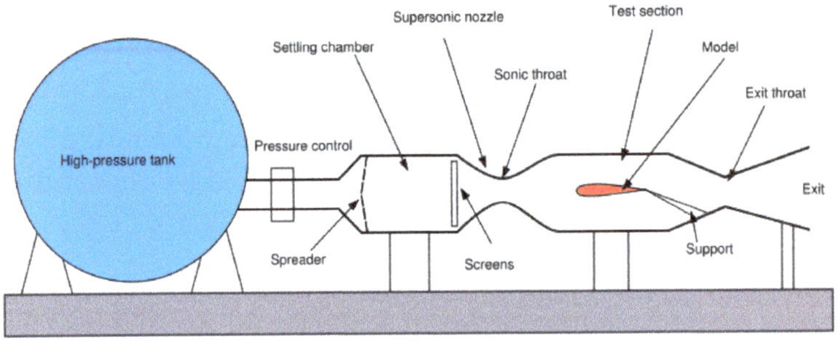

Tunnels such as a Ludwieg tube have short test times (usually less than one second), relatively high Reynolds number, and low power requirements.

Hypersonic Wind Tunnel

NASA Langley's Hypersonic Facilities Complex, 1969

A hypersonic wind tunnel is designed to generate a hypersonic flow field in the working section, thus simulating the typical flow features of this flow regime - including compression shocks and pronounced boundary layer effects, entropy layer and viscous interaction zones and most importantly high total temperatures of the flow. The speed of these tunnels vary from Mach 5 to 15. The power requirement of a wind tunnel increases with the cross section, the flow density and is directly proportional to the third power of the test velocity. Hence installation of a continuous, closed circuit wind tunnel remains a costly affair. The first continuous Mach 7-10 wind tunnel with 1x1 m test section was planned at Kochel am See, Germany during WW II and finally put into operation as 'Tunnel A' in the late 1950s at AEDC Tullahoma, TN, USA for an installed power of 57 MW. In view of these high facility demands, also intermittently operated experimental facilities like blow-down wind tunnels are designed and installed to simulate the hypersonic flow. A hypersonic wind tunnel comprises in flow direction the main components: heater/cooler arrangements, dryer, convergent/divergent nozzle, test section, second throat and diffuser. A blow-down wind tunnel has a low vacuum reservoir at the back end, while a continuously operated, closed circuit wind tunnel has a high power compressor installation instead. Since the temperature drops with the expanding flow, the air inside the test section has the chance of becoming liquefied. For that reason, preheating is particularly critical (the nozzle may require cooling).

Technological Problems

There are several technological problems in designing and constructing a hyper-velocity wind tunnel:

- supply of high temperatures and pressures for times long enough to perform a measurement
- reproduction of equilibrium conditions
- structural damage produced by overheating
- fast instrumentation
- power requirements to run the tunnel

Simulations of a flow at 5.5 km/s, 45 km altitude would require tunnel temperatures of over 9000 K, and a pressure of 3 GPa.

Hot Shot Wind Tunnel

One form of HWT is known as a Gun Tunnel or hot shot tunnel (up to M=27), which can be used for analysis of flows past ballistic missiles, space vehicles in atmospheric entry, and plasma physics or heat transfer at high temperatures. It runs intermittently, but has a very low running time (less than a second). The method of operation is based on a high temperature and pressurized gas (air or nitrogen) produced in an arc-chamber, and a near-vacuum in the remaining part of the tunnel. The arc-chamber can reach several MPa, while pressures in the vacuum chamber can be as low as 0.1 Pa. This means that the pressure ratios of these tunnels are in the order of 10 million. Also, the temperatures of the hot gas are up to 5000 K. The arc chamber is mounted in the gun barrel. The high pressure gas is separated by the vacuum by a diaphragm that breaks down as its resistance is exceeded.

Prior to a test run commencing, a membrane separates the compressed air from the gun barrel breech. A rifle (or similar) is used to rupture the membrane. Compressed air rushes into the breech of the gun barrel, forcing a small projectile to accelerate rapidly down the barrel. Although the projectile is prevented from leaving the barrel, the air in front of the projectile emerges at hypersonic velocity into the working section. Naturally the duration of the test is extremely brief, so high speed instrumentation is required to get any meaningful data.

Classification of Wind Tunnels

Wind tunnels may be classified based on any of the following:

(a) Speed, Mach no

They are classified as of low speed or high speed wind tunnels .In wind tunnel parlance, high speed wind tunnels are those operating at speeds where compressibility effects are important. They are also classified based on the Mach number of operation as subsonic, transonic, supersonic or hypersonic wind tunnels.

(b) Mode of operation (Pressure storage, in-draft or Pressure vacuum type.)

(c) Kind of test section (T.S) - Open, Closed or Semi enclosed

Applications of Wind Tunnels

1. Aerodynamic applications
2. Non-Aero applications in:
   - Civil Engineering
   - Automobile Engineering
   - Calibration of instruments

Model Making, Non-dimensional Parameters

Geometric similarity

One of the most important requirements of models is that there should be geometric similarity between the model and the prototype. By geometric similarity it is meant that ratios of corresponding dimensions in the model and the prototype should be the same.

Dynamic similarity

Equally important as the geometric similarity is the requirement of dynamic similarity. In an actual flight, when the body moves through a medium, forces and moments are generated because of the viscosity of the medium and also due to its inertia, elasticity and gravity. The inertia, viscous, gravity and elastic forces generated on the body in flight can be expressed in terms of fundamental units. The important force ratios can be expressed as non dimensional numbers. For example,

- Reynolds number (Re) = Inertia force/Viscous force
- Mach number = Inertia force/Elastic force
- Froude number = Inertia force/Gravity force

The principle of dynamic similarity is that a scale model under same Reynolds number and Mach number will have forces and moments on it that can be scaled directly. The flow patterns on the full scale body and the model will be exactly similar.

It is not necessary and may not be possible that all the aforesaid non dimensional numbers be simulated simultaneously in any experiment. Depending on the flow regime or the type of experiments, certain non-dimensional parameters are important. For example, in a low speed flow regime, simulation of Reynolds number in the experiments is important to depict the conditions of actual flight. In a high speed flow, simulation of Mach number is significant. It may even be necessary and significant that more than one non dimensional parameter are simulated. The principle of dynamic similarity is applicable in other fields of engineering too. As examples:

Stanton number is simulated in heat transfer experimentation. Stanton no (St) = Heat transferred in to the fluid / Thermal capacity

$$St = \frac{h}{c_p \rho\, v}$$

where h =convective heat transfer coefficient

ρ = density

cp = specific heat at constant pressure

v = velocity

Expressing in terms of non-dimensional parameters,

$$St = \frac{Nu}{\mathrm{Re}^* \, \mathrm{Pr}}$$

$$St = \frac{q}{\rho c_p (T_0 - T_W)}$$

Strouhal number is used in experiments dealing with oscillating flow,

$$S = \frac{f_s I}{u}$$

f_s is vortex shedding frequency,

I is the characteristic length and

v the velocity

Knudsen number $Kn = \frac{\lambda}{l}$ is simulated in low density flows.

In the definition above, λ is the mean free path and l the characteristic dimension

Low Speed Wind Tunnel

Low speed wind tunnels may be of open circuit or closed circuit.

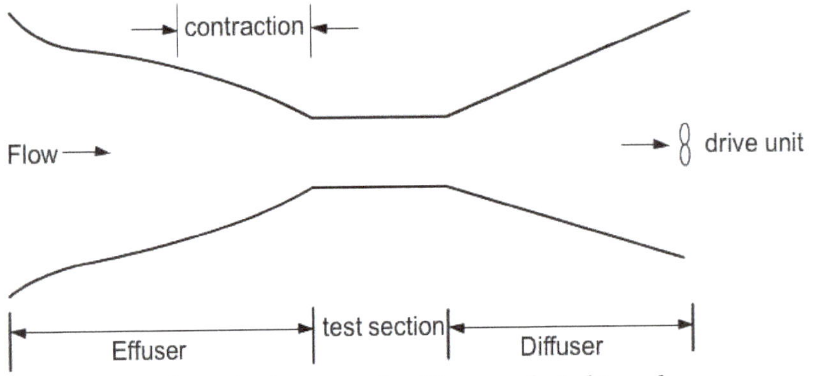

Schematic of an open circuit low speed wind tunnel

Figure above shows an open circuit low speed wind tunnel. After each run, the intake gets air fresh from the atmosphere. The effuser of the wind tunnel is the part of the wind tunnel from the beginning to the entry to the test section. The effuser makes the flow ready for the test section conditions. The test section provides the desired uniform flow conditions along and across the section. It is important that the test section conditions are controllable. Figure below shows a closed circuit wind tunnel. Losses in vorticity, eddies and turbulence are unavoidable in the tunnel circuit. If velocity is large, skin friction and losses due to obstacles will be correspondingly large.

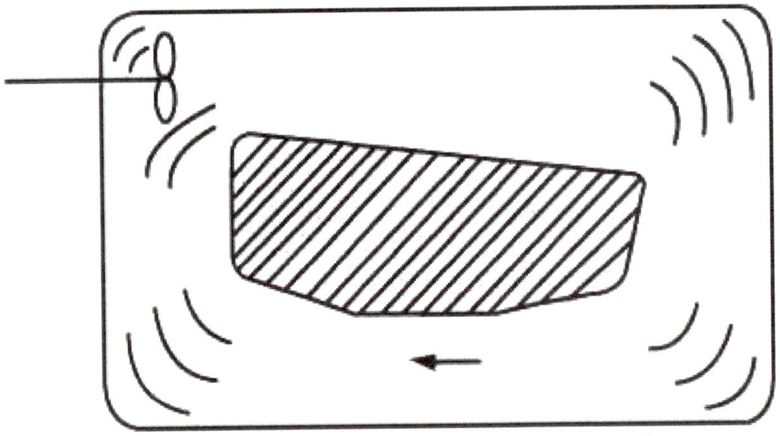

Schematic of an open circuit low speed wind tunnel

Irregularities of Flow In Low Speed Tunnels

1) Spatial non uniformity → Mean velocity not be uniform over a cross section. This is overcome by transferring excess total head from regions of high velocity to those of low velocity.

2) Swirl → Flow may rotate about an axis resulting in variation of direction of flow. Flow straightness and honey combs are used to reduce swirl.

3) Low frequency pulsation → These are surges of mean velocity. Under their influence, time taken for steady conditions becomes excessive. It is difficult to locate the source of such pulsations.

4) Turbulence → Turbulence generates small eddies of varying size and intensity and results in time variations of velocity. Turbulence may be defined as irregular fluctuations of velocity superimposed on mean flow.

In order to quantify turbulence:

Take components of mean velocity as U, V, W

Those of turbulent velocity u, v, w

RMS values $\sqrt{\left(\overline{u}\right)^2}, \sqrt{\left(\overline{v}\right)^2}$ *and* $\sqrt{\left(\overline{w}\right)^2}$ are denoted as u′,v′,w′

Intensity of turbulence $= \frac{u'}{V_0}$ *or* $\frac{v'}{V_0}$ *or* $\frac{w'}{V_0}$ where, V_0 is the mean of U, V, W

Scale of turbulence $L = \int_0^\infty R_y d_y$

where R is the coefficient of co-relation between the longitudinal component of turbulent velocity at A and that at another point B distant y from it.

$$R_y = \frac{\overline{u_A u_B}}{u'_A u'_B}$$

Reduction of Turbulence

Effect of Screens on Turbulence

Use of wire meshes (also called as gauzes or screens) is very common. Screens of very fine mesh size are used. They are kept as far upstream of the test section as possible. Screens are usually made of metal, nylon or polyester. With the use of screens, larger eddies are broken down to smaller ones and the smaller ones decay rapidly. Multiple screens reduce turbulence intensity.

The scale of eddies depends on the flow Re based on wire diameter of the flow through the screens. The eddies are practically absent when the Re is < 40.One of the important reasons for keeping the screens at the beginning of the tunnel circuit is to ensure that they are at the low velocity regions where the Re is the least. Effect of screen on turbulence depends on K, the pressure drop coefficient.

$$K = \frac{p_1 - p_2}{1/2\rho v_1^2}$$

where p_1 and p_2 are values of pressure up and downstream of the screen. K depends also on β, Re, θ. Re the Reynolds number and θ is the flow incidence angle measured from normal to the-screenβ ,is the open area ratio and is defined as:

$\beta = \left(1 - \frac{d}{l}\right)^2$ 'l' and 'd' are marked on the figure.

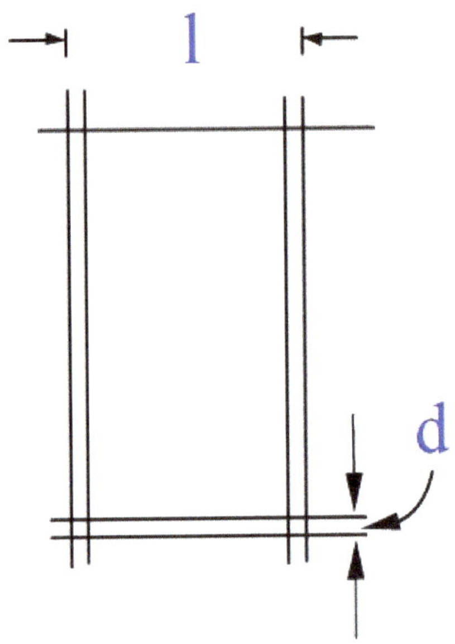

Mesh size of screens

According to Mehta and Bradshaw → If K = 2, turbulence is absent. According to Collar and Batchelor, if $U+u_1$ is the longitudinal velocity far upstream of the screen and $U+u_2$ the corresponding value far downstream, hence

$$\frac{u_2}{u_1} = \frac{2-K}{2+K}$$

so that non-uniformity is removed by a screen whose pressure drop coefficient K is equal to 2.0 and reversed if K is greater than 2.0.

Honey Combs

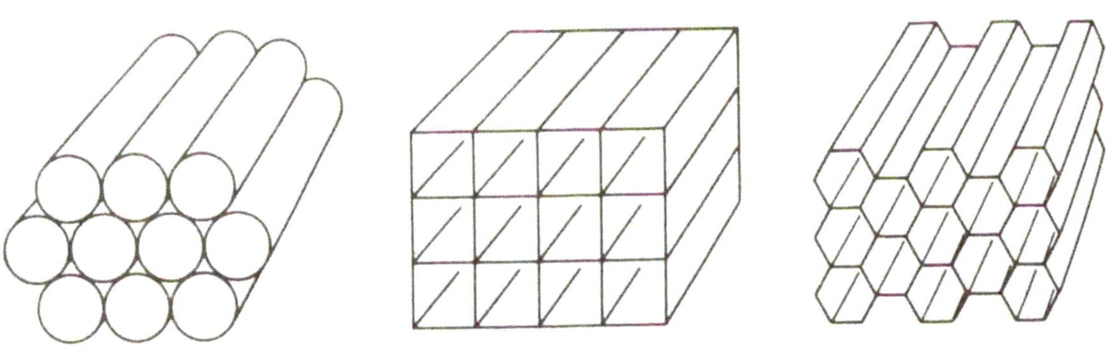

Honeycombs with different cell cross sectional shapes

Honey combs are effective in removing swirl and lateral mean fluctuations. Incidental effect is to reduce turbulence. In order to restrain the boundary layer thickness and transition in to turbulent boundary layer the cell length is usually kept within 5 to 10 times the cell dimension [Cell diameter or cell width].

Wind Tunnel Contractions

Wind tunnel contraction serves a few purposes

(i) Enables velocity to be low at the location of placement of the screens.

(ii) Reduces both mean and fluctuating velocity variations to a smaller fraction of the average velocity.

(iii) Reduces spatial variations of velocity in the wind tunnel cross section

The most important parameter of the contraction is the contraction ratio 'n'. The following one dimensional analysis shows how effective is the contraction in reducing the spatial non uniformity.

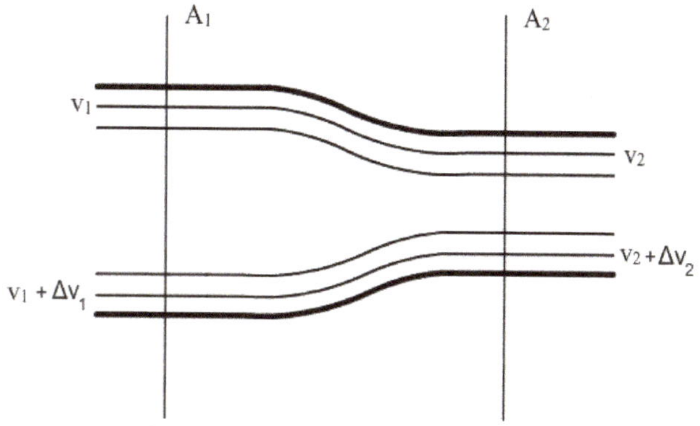

Schematic of a wind tunnel contraction

Referring to the above figure p, v and A represent present pressure, velocity and area respectively.

$a_1 = \frac{\Delta v_1}{v_1}$ is the fractional variation of velocity at the inlet to the contraction.

$a_2 = \frac{\Delta v_2}{v_2}$ is the fractional variation of velocity at the exit of the contraction $n = \frac{A_1}{A_2}$ contraction ratio.

$$p_1 + \frac{\rho v_1^2}{2} = p_2 + \frac{\rho v_2^2}{2}$$

$$p_1 + \frac{\rho}{2}(v_1 + \Delta v_1)^2 = p_2 + \frac{\rho}{2}(v_2 + \Delta v_2)^2$$

$$v_1 \Delta v_1 = v_2 \Delta v_2$$

$$\Delta v_1 = \Delta v_2 \frac{v_2}{v_1}$$

$$a_1 = \Delta v_2 \frac{v_v}{v_1^2} = \Delta v_2 \frac{v_2}{\left(\frac{v_2}{n}\right)^2} = n^2 a_2$$

Diffuser (Automotive)

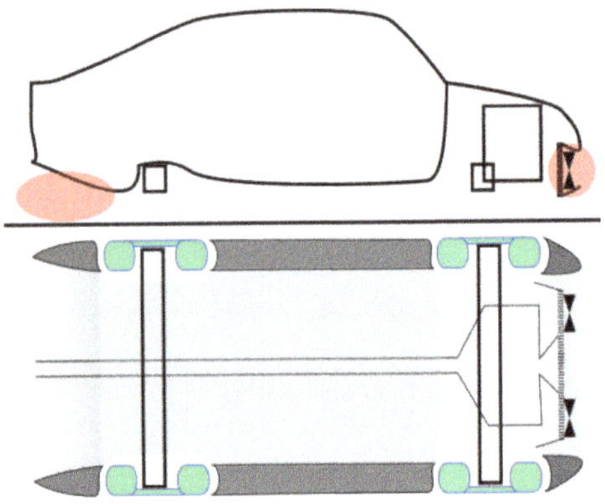

Top: Lateral view; the red circles mark the front air dam/splitter and rear diffuser. Bottom: Underbody view.

A diffuser, in an automotive context, is a shaped section of the car underbody which improves the car's aerodynamic properties by enhancing the transition between the high-velocity airflow underneath the car and the much slower freestream airflow of the ambient atmosphere. It works by providing a space for the underbody airflow to decelerate and expand (in area, density remains constant at the speeds that cars travel) so that it does not cause excessive flow separation and drag, by providing a degree of "wake infill" or more accurately, pressure recovery. The diffuser itself accelerates the flow in front of it, which helps generate downforce.

Overview

When a diffuser is used, the air flows into the underbody from the front of the car, accelerates and reduces pressure. There is a suction peak at the transition of the flat bottom and diffuser. The diffuser then eases this high velocity air back to normal velocity and also helps fill in the area behind the car making the whole underbody a more efficient downforce producing device by reducing drag on the car and increasing downforce. The diffuser also imparts upward momentum to the air which further increases downforce.

Operation

The aft part of a car underbody is where a diffuser is usually located. It works by accelerating the velocity of the airflow underneath the car. The pressure under the car is affected by the diffuser so that it can expand back to ambient in the diffuser, as the car moves through the air. It uses Bernoulli's principle, such that the pressure decreases while the velocity increases. Since the pressure below the car is lower than on the side and above the car, downforce is produced if implemented correctly. The diffuser "drives" the underbody, which produces the downforce. Front diffusers also exist (especially on Le Mans Prototypes or similar cars); however, they generate downforce purely from momentum exchange with the air, as there is nothing ahead of them to drive. A poorly designed front diffuser can create a low pressure region toward the front of the car which slows the air behind it down and reduces the effectiveness of the rest of the underbody. Front diffusers usually route air away from the car so that it doesn't affect the rest of the underbody. The air can be vented through a channel or expelled near the front wheels.

Injecting the exhaust into the rear diffuser can also help extract the air from below the car. The exhaust gasses effectively energize the boundary layer, helping to raise the pressure of the low-pressure, fast-moving airstream back to the ambient atmospheric pressure at the exit of the diffuser. This fast-moving air helps evacuate the diffuser more quickly, which helps drop the pressure at the underbody. However, this makes the diffuser rather sensitive to engine speed. When the driver lifts off the throttle, the exhaust flow is greatly reduced, which makes the diffuser less effective, robbing the vehicle of downforce. Thus, handling is negatively affected.

The car's bodywork also interacts with the flow through the diffuser. In addition to creating downforce, the front wing and nose try to keep "clean air" flowing around, and more importantly under, the car. Clean air under the car prevents flow separation from occurring in the diffuser, which would severely rob its performance. The rear wing also affects the diffuser. When the wing is mounted low and close to the diffuser, the low pressure under the wing helps suck air through the diffuser. Cars, such as the Toyota Eagle MkIII and the Jaguar XJR-14 employed two-tier wings to enhance this effect. One profile was mounted high, in order to hit relatively clean air. The other profile was mounted almost flush with the bodywork behind the chassis. This wing's profile is used to drive the diffuser, creating that low pressure area to help move air from the underbody. According to Hiro Fujimori, aerodynamicist for the Toyota Eagle MkIII project, this bi-plane wing produced 18% more downforce for the same drag as a normal wing. Conversely, equal downforce levels could be attained for significantly reduced drag with this "Red Baron" wing.

Multi-deck Diffusers

In 2009, the Formula 1 grid was embroiled in controversy. The culprit was the so-called *double-decker diffuser* introduced at first by Brawn GP, WilliamsF1, and Toyota Racing, but later put into use by every team. These three teams had exploited a loophole in the rules that allowed for more volume in the diffuser. The rules stated that the diffuser must start at a point aligned with the centerline of the rear wheels. The loophole allowed for holes in the underbody, perpendicular to the reference plane (not visible as a hole when viewed from directly above), that fed a diffuser channel that was above the main diffuser. This greatly increased the available downforce, and was worth about half a second per lap, according to Mike Gascoyne. The teams decided to allow the double-decker diffusers again for 2010. However, for 2011, the Formula 1 Technical Working Group decided to ban multi-deck diffusers.

Splitters

The Porsche RS Spyder Evo uses a front splitter combined with dive planes

As the front of the car slows down the air without a diffuser, this is the ideal place for an inlet. A splitter is commonly used here, serving to increase the amount of downforce at the front of the car. The airstream is brought to stagnation above the splitter by an air dam, causing an area of high pressure. Below the splitter, the air is redirected away from the stagnation zone and is accelerated, causing the pressure to drop. This, combined with the high pressure over the splitter, creates downforce. The larger the area of the splitter, the more downforce is generated. In most closed-wheel race cars, the underside of the splitter smoothly integrates with the undertray, creating one large flat plane that is driven by the rear diffuser. Some race cars, such as the Toyota GT-One, actually use an additional diffuser immediately behind the splitter to help create more downforce. The air extracted by this diffuser is exhausted through vents in the sidepods or above the car around the cockpit.

The diffuser in the wind tunnel serves the purpose of salvaging the kinetic energy of flow in the test section as pressure energy. A well designed diffuser does this efficiently. In subsonic wind tunnels, the diffusers are diverging passages with a semi divergence angle of about 7.5 to 8.0 degrees. The Bernoulli's equation written in differential form in the context of a diffuser is as follows:

$$d\left(\frac{v^2}{2}\right)+\frac{dp}{\rho}=0$$

This implies that for a decrease of kinetic energy $d\left(\frac{v^2}{2}\right)$ per unit mass, there is a corresponding increase in pressure energy. The pressure gradient in a subsonic diverging passage is adverse. It is difficult to avoid boundary layer thickening and flow separation. Hence, the conversion of kinetic energy into pressure energy is never fully efficient.

The efficiency of the diffuser is best understood in physical terms when the efficiency term is included in the Bernoulli's equation as below:

$\eta_0\left(\frac{dv^2}{2}\right)+\frac{dp}{\rho}=0$ where η_D is the diffuser efficiency

Pressure changes in expanding passages may be examined by referring to the figure below to elucidate the statement above.

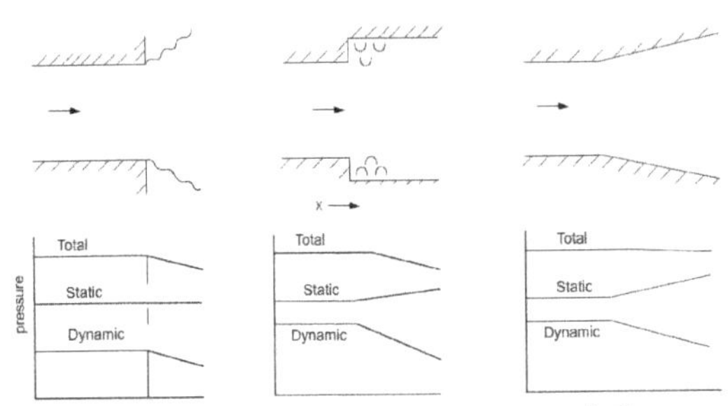

Exit pressure profile of jet through different passages

There are two definitions of diffuser efficiency

a) Polytrophic efficiency η_D

$$\eta_D = \frac{p_2 - p_1}{\frac{1}{2}\rho v_1^2 - \frac{1}{2}\rho v_2^2}$$

Use continuity equation to write the above equation as

$$\eta_D = \frac{p_2 - p_1}{\frac{1}{2}\rho v_1^2\left[1-\left(\frac{A_1}{A_2}\right)^2\right]}$$

where subscripts '1' and '2' refer to conditions at the entry and exit of the diffuser.

As explained before, the equation is indicative that from kinetic energy to pressure energy it is not fully converted. Loss of total head in the diffuser action:

$$\Delta H = \left(\frac{1}{2}\rho v_1^2 - \frac{1}{2}\rho v_2^2\right) - (p_2 - p_1)$$

$$\eta_D = 1 - \frac{\Delta H}{\frac{1}{2}\rho v_1^2\left[1-\left(\frac{A_1}{A_2}\right)^2\right]}$$

(a) *Isentropic efficiency*

Isentropic Efficiency (η_σ) is defined as the ratio of

$$\eta_\sigma = \frac{\textit{Kinetic energy which would have to be transformed to produce the observed pressure recovery}}{\textit{Kenetic energy actually transformed}}$$

$\left(\frac{p}{\rho^\gamma}\right)$ = constant for an isentropic process

KE to be transformed to raise pressure from p_1 *to* $p_2 = \int_{P_1}^{P_2} \frac{dp}{\rho}$

$$\int_{P_1}^{P_2} \frac{dp}{\rho} = \frac{\gamma}{\gamma-1}\frac{p_1}{p_1}\left[\left(\frac{p_2}{p_1}\right)^{\frac{\gamma-1}{\gamma}} - 1\right]$$

$$\eta_\sigma = \frac{\gamma}{\gamma-1}\frac{p_1}{p_1}\left[\left(\frac{p_2}{p_1}\right)^{\frac{\gamma-1}{\gamma}} - 1\right]$$

$$\frac{p_2}{p_1} = \left[\frac{(\gamma-1)M_1^2}{2}\eta_\sigma + 1\right]^{\gamma/\gamma-1}$$

$$\frac{p_2}{H} = \frac{p_2}{p_{01}} = \left[\frac{2}{2+(\gamma-1)M_1^2}\right]$$

Overall pressure ratio $\frac{p_{01}}{p_2}=\frac{H}{p_2}$. *Here p_2 corresponds to the pressure at the exit of the diffuser* and H represents the stagnation pressure p_{01} at the entry to the wind tunnel.

$$\frac{H}{p_1}\times\frac{p_1}{p_2}=\left[\frac{2+(\gamma-1)M_1^2}{2+((\gamma-1)M_1^2\eta_\sigma}\right]^{\gamma/\gamma-1}$$

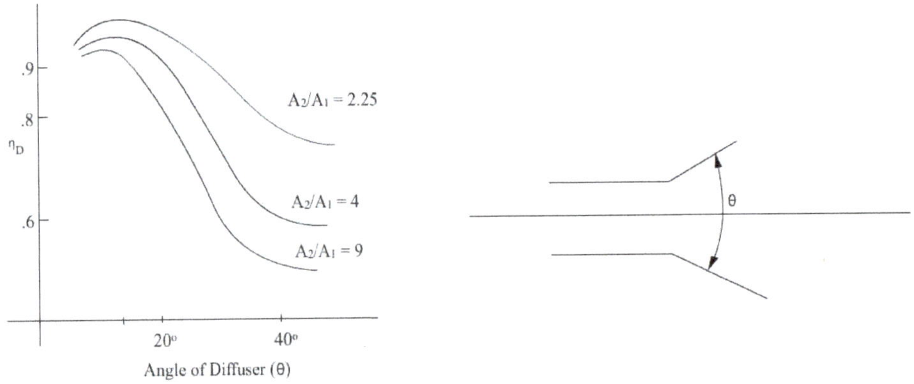

Diffuser efficiency as a function of diffuser angle

Losses in the Wind Tunnel Circuit

Losses are due to:

- Inefficiency of drive unit
- Skin friction, separation etc
- Loss of kinetic energy at the diffuser exit
- Shocks in the case of supersonic wind -tunnels

Losses due to skin friction

Local coefficient of skin friction $=\frac{Frictinal\ force}{\frac{1}{2}\rho v^2 A'}$

where, A' is the surface area of the solid boundary which is subjected to frictional force.

$$\Delta H=\int C_1\frac{1}{2}\rho v^2\frac{L}{A}dS$$

Integral is taken over the length of the duct.

L is the perimeter, ds is an element of length in the direction of flow.

A is the cross sectional area and

Cf depends on nature of the boundary layer, Reynolds number and on the surface nature.

Losses Due to Resistance in the Wind Tunnel Circuit

Corner vanes, gauzes and screen offer resistance

$$\text{Resistance Coefficient C} = \frac{\text{Resistance force}}{\frac{1}{2}\rho v^2 A} = \frac{\Delta H}{\frac{1}{2}\rho v^2}$$

vA = constant

$$\text{v} = \frac{\text{constant}}{\text{A}}$$

$\Delta H \, a \frac{C}{A^2}$, which shows that the power loss is less for large areas.

Power Requirements – Power Economy

Power Factor is defined as

$$\lambda = \frac{\textit{Power Input}}{\textit{Rate of flow of kinetic energy in the test section}} = \frac{P}{\frac{1}{2}\rho v^3 A}$$

where P is the power input. Of the power P, only ηP is communicated to the air stream where η is the efficiency of the drive unit (fan efficiency).

$$\lambda = \frac{\Sigma\, losses}{\eta \frac{1}{2}\rho v^3 A} \qquad \eta^P = \text{losses in the wind tunnel.}$$

Reciprocal of the power factor is an alternative measure of the efficiency of the system

Power Economy

$$P = \lambda \frac{1}{2}\rho v^3 A$$

$$= \lambda \frac{1}{2}\rho v^2 v A$$

$$\text{Re} = \frac{\rho v c}{\mu}, M = \frac{v}{a}$$

$$a^2 = \gamma RT = \frac{\gamma p}{\rho}$$

$$\rho = \frac{\gamma p}{a^2}$$

$$a^2 = \gamma \frac{R^*}{M} T$$

$$v = \frac{\text{Re}\,\mu}{\rho c}$$

substituting from above

$$P = \lambda \frac{1}{2} \frac{\mathrm{Re}^2 \mu^2}{\rho^2 c^2} AMa\rho$$

$$= \frac{1}{2} \lambda \frac{A}{c^2} \mathrm{Re}^2 M \frac{\mu^2 a^3}{\gamma p}$$

The ratio $\frac{A}{c^2}$ = the tunnel interference, is constant.

The power factor depends largely on the geometry of the tunnel and on the Mach and Reynolds numbers.

Power Economy by Pressurization

For tunnels of similar geometry, operating at given Re and M.

$p\alpha \frac{\mu^2 a^3}{\gamma p}$ Expressing in terms of stagnation properties,

$\alpha \frac{\mu_0^2 a_0^3}{\gamma p_0}$

$p\,\alpha \frac{1}{p_0}$ Power is inversely proportional to p_0.

Objections to power economy by pressurization

Aerodynamic forces on the model are proportional to $\frac{1}{2}\rho v^2 = \left[\frac{1}{2}\gamma p M^2\right]$

which for given M is proportional to p_0.

Power Economy by Choice of Working Fluid

For a given pressure $\rightarrow p\,\alpha \frac{\mu_0^2 a_0^3}{\gamma}$

At constant temperature $a_0^3\,\alpha \left(\frac{\gamma}{\overline{M}}\right)^{3/2}$

$$\alpha^3 = \left(\frac{\gamma R^* T}{\overline{M}}\right)^{3/2}$$

μ_0 falls as the number of atoms in a molecule increases.

$\gamma = \frac{2n+3}{2n+1}$ where n is the number of atoms in the molecule.

γ falls when n increases

$p\,\alpha \frac{\mu_0^2 \gamma^{1/2}}{\overline{M}^{3/2}} \rightarrow$ Power economy can be achieved by working fluid of higher molecular weight

Limitations

→ Complete dynamic similarity can be got with only the same γ .

→ Boiling point of higher molecular weight fluids is high. Hence, to keep the working fluid in gaseous state, the temperature should be high.

Table: Power required with a different working substance

T_o = 288 K

Fluid	Boiling Point K			Power relative to air
Air	90	ao SF6 / aoAir = 0.395µ	µoSF6/ µoAir =0.529	0.020
SF6	222			

The table above shows that the power required when SF6 is used as the working fluid is only 1/50 of that while using air.

Power Economy by Reduction in Stagnation Temperature

For a given working fluid, po, λ , M and Re.

$$P \,\alpha \mu_0^2 a_0^3$$

$$a_0 \;\alpha\; T_0^{1/2}$$

$$\mu_0 \;\alpha \frac{T_0^{3/2}}{T_{Oref}+C} \qquad C \textit{ is a } \text{constant}.$$

$$P \,\alpha \frac{T_0^{9/2}}{\left(T_{Oref}+C\right)^2}$$

Objections

- Reducing stagnation temperature brings down the flow temperature too. Boiling point is one consideration.
- Severe metallurgical problems at low temperatures.

High Speed Wind Tunnels

Definition of High Speed

When compressibility effects are pre dominant the flow is generally said to be of high speed. A lower limit is approximately M=0.5. Power requirements vary as cube of velocity in the wind tunnel. This does not hold into the high speed regime exactly. Because of large power requirements, high speed wind tunnels are of the intermittent type.

Types of High Speed Wind Tunnels

- Continuous (for all speed ranges)
- Intermittent
    - Blowdown (fig. a) } → M > 0.5 < 5.0
    - Indraft (fig. b)
    - Intermittent pressure vacuum tunnel for M > 5

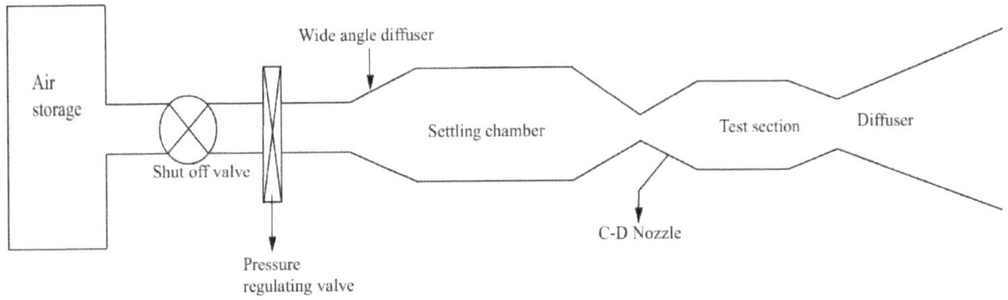

(a) Pressure driven blow down supersonic wind tunnel

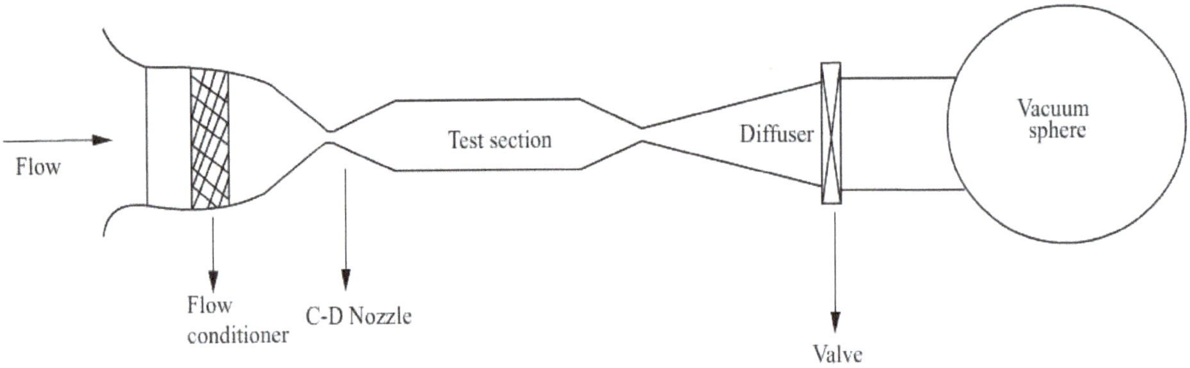

(b) Indraft type wind tunnel

Table (a) Comparison between Indraft and Pressure driven wind tunnels

Indraft wind tunnels	**Pressure driven wind tunnels**
Stagnation temperature at supply condition is constant during a run. So also is total pressure. No fluctuations as those generated by a pressure regulator.	Reynolds number can be varied at a particular Mach no.
No possible contamination such as that due to oil.	Cost is much less than of an indraft tunnel.
Vacuum is safer to handle than pressure.	
Pressure regulators are not needed.	

The wind tunnels described above can be converted as continuous tunnels. The comparison between blow down and continuous wind tunnels are as given in Table (b).

Table (b) Comparison between intermittent and continuous wind tunnels

Intermittent(blow down) wind tunnels	Continuous wind tunnels
Simple to design and less costly	More in control of conditions and return to a given test condition with more accuracy.
A single drive may run several Tunnels	Check points are easily obtained No panic of rapid testing
Model testing is more convenient	Test conditions can be held constant for a longer time.
Extra power is available to start	
Failure of model will not result in tunnel damage	

Supersonic Wind Tunnels

The nozzle regulates the speed of air entering the test section (T.S) of the wind tunnel so that the desired Mach number is established. Mach number is uniquely determined by the area ratio of the nozzle. A well designed nozzle makes the flow parameters uniform across the cross section. The design of a suitably shaped nozzle contour to obtain the desired uniform flow at the nozzle exit is based on the method of characteristics.

Dynamic Pressure

In incompressible fluid dynamics dynamic pressure (indicated with q, or Q, and sometimes called velocity pressure) is the quantity defined by:

$$q = \tfrac{1}{2}\rho u^2,$$

where (using SI units):

q	= dynamic pressure in pascals,
ρ	= fluid density in kg/m³ (e.g. density of water),
u	= flow speed in m/s.

Physical Meaning

Dynamic pressure is the kinetic energy per unit volume of a fluid particle. Dynamic pressure is in fact one of the terms of Bernoulli's equation, which can be derived from the conservation of energy for a fluid in motion. In simplified cases, the dynamic pressure is equal to the difference between the stagnation pressure and the static pressure.

Another important aspect of dynamic pressure is that, as dimensional analysis shows, the aerodynamic stress (i.e. stress within a structure subject to aerodynamic forces) experienced by an aircraft travelling at speed v is proportional to the air density and square of v, i.e. proportional to q. Therefore, by looking at the variation of q during flight, it is possible to determine how the stress will vary and in particular when it will reach its maximum value. The point of maximum aerodynamic load is often referred to as *max Q* and it is a critical parameter in many applications, such as during spacecraft launch.

Uses

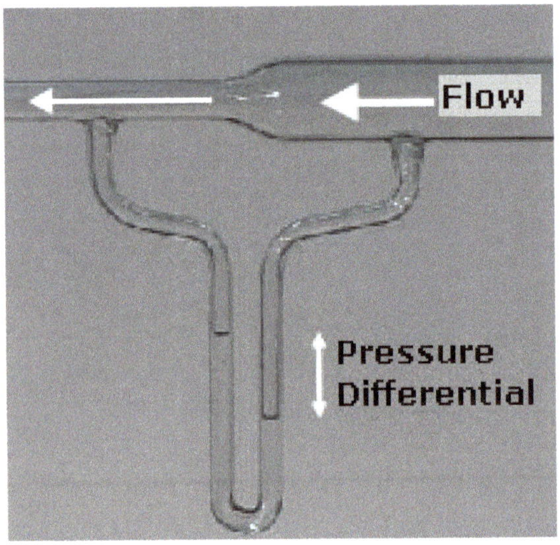

A flow of air through a venturi meter, showing the columns connected in a U-shape (a manometer) and partially filled with water. The meter is "read" as a differential pressure head in cm or inches of water and is equivalent to the difference in velocity head.

The dynamic pressure, along with the static pressure and the pressure due to elevation, is used in Bernoulli's principle as an energy balance on a closed system. The three terms are used to define the state of a closed system of an incompressible, constant-density fluid.

If we were to divide the dynamic pressure by the product of fluid density and acceleration due to gravity, g, the result is called velocity head, which is used in head equations like the one used for pressure head and hydraulic head. In a venturi flow meter, the *differential pressure head* can be used to calculate the *differential velocity head*, which are equivalent in the adjacent picture. An alternative to *velocity head* is *dynamic head.*

Compressible Flow

Many authors define *dynamic pressure* only for incompressible flows. (For compressible flows, these authors use the concept of impact pressure.) However, the definition of *dynamic pressure* can be extended to include compressible flows.

If the fluid in question can be considered an ideal gas (which is generally the case for air), the dynamic pressure can be expressed as a function of fluid pressure and Mach number.

By applying the ideal gas law:

$$p_s = \rho_m RT,$$

the definition of the speed of sound a and of Mach number M :

$$a = \sqrt{\frac{\gamma RT}{m_m}} \text{ and } M = \frac{u}{a},$$

and also $q = \frac{1}{2}\rho u^2$, dynamic pressure can be rewritten as:

$$q = \tfrac{1}{2}\gamma p_s M^2,$$

where (using SI units):

P_s	= static pressure in Pascals, Is also the basic SI unit of Pressure
ρ_m	= molar density of the ideal gas in mol/m³
m_m	= mass of a mole of the ideal gas in kg/mol
$\rho = \rho_m m_m$	= density of the ideal gas in kg/m³
	= gas constant (8.3144 J/(mol·K)),
T	= absolute temperature in kelvins (K),
M	= Mach number (non-dimensional),
γ	= ratio of specific heats (non-dimensional) (1.4 for air at sea level conditions),
u	= flow speed in m/s,
a	= speed of sound in m/s

Mass Flow Rate

In physics and engineering, mass flow rate is the mass of a substance which passes per unit of time. Its unit is kilogram per second in SI units, and slug per second or pound per second in US customary units. The common symbol is $\dot{m}$, although sometimes μ (Greek lowercase mu) is used.

Sometimes, mass flow rate is termed *mass flux* or *mass current*, see for example *Fluid Mechanics, Schaum's et al.* In this article, the (more intuitive) definition is used.

Mass flow rate is defined by the limit:

$$\dot{m} = \lim_{\Delta t \to 0} \frac{\Delta m}{\Delta t} = \frac{dm}{dt}$$

i.e. the flow of mass *m* through a surface per unit time *t*.

The overdot on the m is Newton's notation for a time derivative. Since mass is a scalar quantity, the mass flow rate (the time derivative of mass) is also a scalar quantity. The change in mass is the amount that flows *after* crossing the boundary for some time duration, not the initial amount of mass at the boundary minus the final amount at the boundary, since the change in mass flowing through the area would be zero for steady flow.

Alternative Equations

Mass flow rate can also be calculated by:

$$\dot{m} = \rho \cdot \dot{V} = \rho \cdot \mathbf{v} \cdot \mathbf{A} = \mathbf{j}_{\mathrm{m}} \cdot \mathbf{A}$$

where:

- $\dot{V}$ or Q = Volume flow rate,
- ρ = mass density of the fluid,
- $\mathbf{v}$ = Flow velocity of the mass elements,
- $\mathbf{A}$ = cross-sectional vector area/surface,
- $\mathbf{j}_{\mathrm{m}}$ = mass flux.

The above equation is only true for a flat, plane area. In general, including cases where the area is curved, the equation becomes a surface integral:

$$\dot{m} = \iint_A \rho \mathbf{v} \cdot d\mathbf{A} = \iint_A \mathbf{j}_{\mathrm{m}} \cdot d\mathbf{A}$$

The area required to calculate the mass flow rate is real or imaginary, flat or curved, either as a cross-sectional area or a surface. E.g. for substances passing through a filter or a membrane, the real surface is the (generally curved) surface area of the filter, macroscopically - ignoring the area spanned by the holes in the filter/membrane. The spaces would be cross-sectional areas. For liquids passing through a pipe, the area is the cross-section of the pipe, at the section considered. The vector area is a combination of the magnitude of the area through which the mass passes through, A, and a unit vector normal to the area, $\hat{\mathbf{n}}$. The relation is $\mathbf{A} = A\hat{\mathbf{n}}$..

The reason for the dot product is as follows. The only mass flowing *through* the cross-section is the amount normal to the area, i.e. parallel to the unit normal. This amount is:

$$\dot{m} = \rho v A \cos\theta$$

where θ is the angle between the unit normal $\hat{\mathbf{n}}$ and the velocity of mass elements. The amount passing through the cross-section is reduced by the factor $\cos\theta$,, as θ increases less mass passes through. All mass which passes in tangential directions to the area, that is perpendicular to the unit normal, *doesn't* actually pass *through* the area, so the mass passing through the area is zero. This occurs when $\theta = \pi/2$:

$$\dot{m} = \rho v A \cos(\pi / 2) = 0$$

These results are equivalent to the equation containing the dot product. Sometimes these equations are used to define the mass flow rate.

Considering flow through porous media, a special quantity, superficial mass flow rate, can be introduced. It is related with superficial velocity, v_s, with the following relationship:

$$\dot{m}_s = v_s \cdot \rho = \dot{m} / A$$

The quantity can be used in particle Reynolds number or mass transfer coefficient calculation for fixed and fluidized bed systems.

Usage

In the elementary form of the continuity equation for mass, in Hydrodynamics:

$$\rho_1 \mathbf{v}_1 \cdot \mathbf{A}_1 = \rho_2 \mathbf{v}_2 \cdot \mathbf{A}_2$$

In elementary classical mechanics, mass flow rate is encountered when dealing with objects of variable mass, such as a rocket ejecting spent fuel. Often, descriptions of such objects erroneously invoke Newton's second law $\mathbf{F} = \mathrm{d}(m\mathbf{v})/\mathrm{d}t$ by treating both the mass m and the velocity $\mathbf{v}$ as time-dependent and then applying the derivative product rule. A correct description of such an object requires the application of Newton's second law to the entire, constant-mass system consisting of both the object and its ejected mass.

Analogous Quantities

In hydrodynamics, mass flow rate is the rate of flow of mass. In electricity, the rate of flow of charge is electric current.

Reynolds Number

The plume from this candle flame goes from laminar to turbulent. The Reynolds number can be used to predict where this transition will take place.

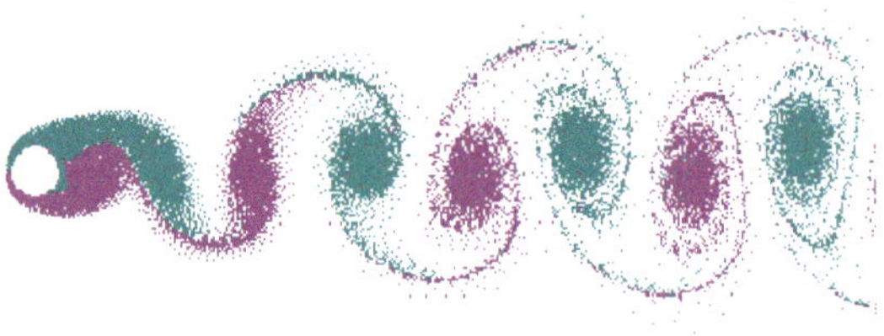

A vortex street around a cylinder. This can occur around cylinders and spheres, for any fluid, cylinder size and fluid speed, provided that the flow has a Reynolds number between roughly 40 and 1000.

Sir George Stokes, introduced Reynolds numbers

Osborne Reynolds popularised the concept

The Reynolds number (Re) is an important dimensionless quantity in fluid mechanics used to help predict flow patterns in different fluid flow situations. It has wide applications, ranging from liquid flow in a pipe to the passage of air over an aircraft wing. The Reynolds Number is used to predict the transition from laminar to turbulent of flow, and used in the scaling of similar but different-sized flow situations, such as between an aircraft model in a wind tunnel and the full size version. The predictions of onset of turbulence and the ability to calculate scaling effects can be used to help predict fluid behaviour on a larger scale, such as in local or global air or water movement and thereby the associated meteorological and climatological effects.

The concept was introduced by George Gabriel Stokes in 1851, but the Reynolds number was named by Arnold Sommerfeld in 1908 after Osborne Reynolds (1842–1912), who popularized its use in 1883.

Definition

The Reynolds number is the ratio of inertial forces to viscous forces within a fluid which is subjected to relative internal movement due to different fluid velocities, in what is known as a boundary layer in the case of a bounding surface such as the interior of a pipe. A similar effect is created by the introduction of a stream of higher velocity fluid, such as the hot gases from a flame in air. This relative movement generates fluid friction, which is a factor in developing turbulent flow. Counteracting this effect is the viscosity of the fluid, which as it increases, progressively inhibits turbulence, as more kinetic energy is absorbed by a more viscous fluid. The Reynolds number quantifies the relative importance of these two types of forces for given flow conditions, and is a guide to when turbulent flow will occur in a particular situation.

This ability to predict the onset of turbulent flow is an important design tool for equipment such as piping systems or aircraft wings, but the Reynolds number is also used in scaling of fluid dynamics problems, and is used to determine dynamic similitude between two different cases of fluid flow, such as between a model aircraft, and its full size version. Such scaling is not linear and the application of Reynolds numbers to both situations allows scaling factors to be developed.

With respect to laminar and turbulent flow regimes:

- laminar flow occurs at low Reynolds numbers, where viscous forces are dominant, and is characterized by smooth, constant fluid motion;
- turbulent flow occurs at high Reynolds numbers and is dominated by inertial forces, which tend to produce chaotic eddies, vortices and other flow instabilities.

The Reynolds number is defined as

$$\mathrm{Re} = \frac{\rho u L}{\mu} = \frac{u L}{\nu}$$

where:

- ρ is the density of the fluid (SI units: kg/m^3)
- u is the velocity of the fluid with respect to the object (m/s)

- L is a characteristic linear dimension (m)
- μ is the dynamic viscosity of the fluid (Pa·s or N·s/m^2 or kg/(m·s))
- ν is the kinematic viscosity of the fluid (m^2/s).

The Reynolds number can be defined for several different situations where a fluid is in relative motion to a surface.[n 1] These definitions generally include the fluid properties of density and viscosity, plus a velocity and a characteristic length or characteristic dimension (L in the above equation). This dimension is a matter of convention – for example radius and diameter are equally valid to describe spheres or circles, but one is chosen by convention. For aircraft or ships, the length or width can be used. For flow in a pipe or a sphere moving in a fluid the internal diameter is generally used today. Other shapes such as rectangular pipes or non-spherical objects have an *equivalent diameter* defined. For fluids of variable density such as compressible gases or fluids of variable viscosity such as non-Newtonian fluids, special rules apply. The velocity may also be a matter of convention in some circumstances, notably stirred vessels.

In practice, matching the Reynolds number is not on its own sufficient to guarantee similitude. Fluid flow is generally chaotic, and very small changes to shape and surface roughness can result in very different flows. Nevertheless, Reynolds numbers are a very important guide and are widely used.

History

Osborne Reynolds' apparatus of 1883 demonstrating the onset of turbulent flow. The apparatus is still at the University of Manchester.

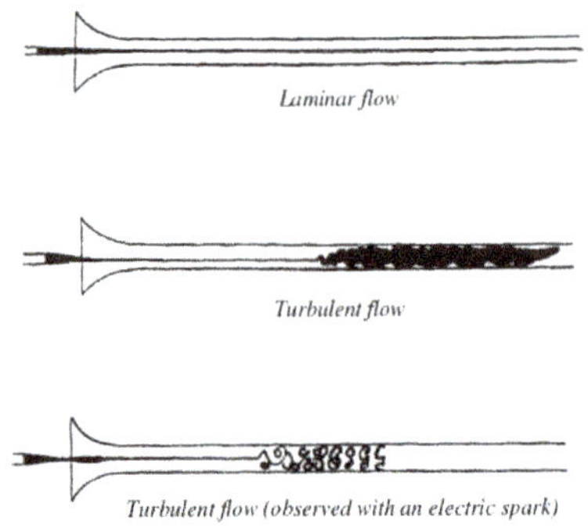

Diagram from Reynolds' 1883 paper showing onset of turbulent flow.

Osborne Reynolds famously studied the conditions in which the flow of fluid in pipes transitioned from laminar flow to turbulent flow. In his 1883 paper Reynolds described the transition from laminar to turbulent flow in a classic experiment in which he examined the behaviour of water flow under different flow velocities using a small stream of dyed water introduced into the centre of clear water flow in a larger pipe.

The larger pipe was glass so the behaviour of the layer of the dyed stream could be observed, and at the end of this pipe there was a flow control valve used to vary the water velocity inside the tube. When the velocity was low, the dyed layer remained distinct through the entire length of the large tube. When the velocity was increased, the layer broke up at a given point and diffused throughout the fluid's cross-section. The point at which this happened was the transition point from laminar to turbulent flow.

From these experiments came the dimensionless Reynolds number for dynamic similarity—the ratio of inertial forces to viscous forces. Reynolds also proposed what is now known as the Reynolds-averaging of turbulent flows, where quantities such as velocity are expressed as the sum of mean and fluctuating components. Such averaging allows for 'bulk' description of turbulent flow, for example using the Reynolds-averaged Navier–Stokes equations.

Derivation

The form of the Reynolds number can be derived as follows

$$\mathrm{Re} = \frac{\text{inertial forces}}{\text{viscous forces}} = \frac{(\text{mass})(\text{acceleration})}{(\text{dynamic viscosity})\left(\frac{\text{velocity}}{\text{distance}}\right)(\text{area})} = \frac{(\rho L^3)\left(\frac{v}{t}\right)}{\mu\left(\frac{v}{L}\right)L^2} = \frac{(\rho L^3)\left(\frac{1}{t}\right)}{\mu\left(\frac{1}{L}\right)L^2} = \frac{(\rho L^2)\left(\frac{1}{t}\right)}{\mu} = \frac{(\rho)\left(\frac{L}{t}\right)(L)}{\mu} = \frac{\rho v L}{\mu} = \frac{vL}{\nu}$$

where:

- v is the maximum velocity of the object relative to the fluid (SI units: m/s)
- L is a characteristic linear dimension, (travelled length of the fluid; hydraulic diameter when dealing with river systems) (m)
- t denotes the time
- μ is the dynamic viscosity of the fluid (Pa·s or N·s/m² or kg/(m.s))
- ν (nu) is the kinematic viscosity ($\nu = \mu/\rho$) (m²/s)
- ρ is the density of the fluid (kg/m³).

Note that multiplying the Reynolds number by **Lv**/**Lv** yields $\frac{\rho v^2 L^2}{\mu v L}$, which is the ratio of the inertial forces to the viscous forces. It could also be considered the ratio of the total momentum transfer to the molecular momentum transfer.

Flow in a Pipe

For flow in a pipe or tube, the Reynolds number is generally defined as:

$$\mathrm{Re} = \frac{\rho v D_{\mathrm{H}}}{\mu} = \frac{v D_{\mathrm{H}}}{\nu} = \frac{Q D_{\mathrm{H}}}{\nu A}$$

where:

- D_{H} is the hydraulic diameter of the pipe (the inside diameter if the pipe is circular) (m).
- Q is the volumetric flow rate (m^3/s).
- A is the pipe's *cross-sectional* area (m^2).
- υ is the mean velocity of the fluid (m/s).
- μ is the dynamic viscosity of the fluid ($Pa{\cdot}s = N{\cdot}s/m^2 = kg/(m^*s)$).
- ν (nu) is the kinematic viscosity ($\nu = \mu/\rho$) (m^2/s).
- ρ is the density of the fluid (kg/m^3).

For shapes such as squares, rectangular or annular ducts where the height and width are comparable, the characteristical dimension for internal flow situations is taken to be the hydraulic diameter, D_{H}, defined as:

$$D_{\mathrm{H}} = \frac{4A}{P},$$

where A is the cross-sectional area and P is the wetted perimeter. The wetted perimeter for a channel is the total perimeter of all channel walls that are in contact with the flow. This means the length of the channel exposed to air is *not* included in the wetted perimeter.

For a circular pipe, the hydraulic diameter is exactly equal to the inside pipe diameter, D. That is,

$$D_{\mathrm{H}} = D.$$

For an annular duct, such as the outer channel in a tube-in-tube heat exchanger, the hydraulic diameter can be shown algebraically to reduce to

$$D_{\mathrm{H,annulus}} = D_{\mathrm{o}} - D_{\mathrm{i}}$$

where

D_{o} is the inside diameter of the outside pipe, and

D_{i} is the outside diameter of the inside pipe.

For calculations involving flow in non-circular ducts, the hydraulic diameter can be substituted for the diameter of a circular duct, with reasonable accuracy, if the aspect ratio AR of the duct cross-section remains in the range $1/4 < \mathrm{AR} < 4$.

Laminar-turbulent Transition

In boundary layer flow over a flat plate, experiments confirm that, after a certain length of flow, a laminar boundary layer will become unstable and turbulent. This instability occurs across different scales and with different fluids, usually when $Re_x \approx 5\times10^5$, where x is the distance from the leading edge of the flat plate, and the flow velocity is the freestream velocity of the fluid outside the boundary layer.

For flow in a pipe of diameter D, experimental observations show that for "fully developed" flow,laminar flow occurs when $Re_D < 2000$ and turbulent flow occurs when $Re_D > 4000$. In the interval between 2000 and 4000, laminar and turbulent flows are possible and are called "transition" flows, depending on other factors, such as pipe roughness and flow uniformity. This result is generalized to non-circular channels using the hydraulic diameter, allowing a transition Reynolds number to be calculated for other shapes of channel.

These transition Reynolds numbers are also called *critical Reynolds numbers*, and were studied by Osborne Reynolds around 1895. The critical Reynolds number is different for every geometry.

Flow in a Wide Duct

For a fluid moving between two plane parallel surfaces—where the width is much greater than the space between the plates—then the characteristic dimension is equal to the distance between the plates.

Flow in an Open Channel

For flow of liquid with a free surface, the hydraulic radius must be determined. This is the cross-sectional area of the channel divided by the wetted perimeter. For a semi-circular channel, it is half the radius. For a rectangular channel, the hydraulic radius is the cross-sectional area divided by the wetted perimeter. Some texts then use a characteristic dimension that is four times the hydraulic radius, chosen because it gives the same value of Re for the onset of turbulence as in pipe flow, while others use the hydraulic radius as the characteristic length-scale with consequently different values of Re for transition and turbulent flow.

Flow Around Airfoils

Reynolds numbers are used in airfoil design to (among other things) manage "Scale Effect" when computing/comparing characteristics (a tiny wing, scaled to be huge, will perform differently). Fluid dynamicists define the chord Reynolds number, R, like this: $R = Vc/\nu$ where V is the flight speed, c is the chord length, and ν is the kinematic viscosity of the fluid in which the airfoil operates, which is 1.460×10^{-5} m²/s for the atmosphere at sea level. In some special studies a characteristic length other than chord may be used; rare is the "span Reynolds number" which is not to be confused with spanwise stations on a wing where chord is still used.

Object in a Fluid

The Reynolds number for an object moving in a fluid, called the particle Reynolds number and often denoted Re_p. It characterizes the nature of the surrounding flow and its fall velocity.

In Viscous Fluids

Where the viscosity is naturally high, such as polymer solutions and polymer melts, flow is normally laminar. The Reynolds number is very small and Stokes' Law can be used to measure the viscosity of the fluid. Spheres are allowed to fall through the fluid and they reach the terminal velocity quickly, from which the viscosity can be determined.

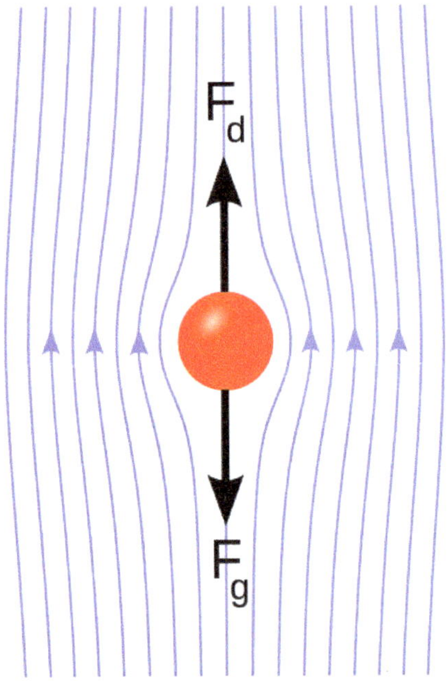

Creeping flow past a falling sphere: streamlines, drag force F_d and force by gravity F_g.

The laminar flow of polymer solutions is exploited by animals such as fish and dolphins, who exude viscous solutions from their skin to aid flow over their bodies while swimming. It has been used in yacht racing by owners who want to gain a speed advantage by pumping a polymer solution such as low molecular weight polyoxyethylene in water, over the wetted surface of the hull.

It is, however, a problem for mixing of polymers, because turbulence is needed to distribute fine filler through the material. Inventions such as the "cavity transfer mixer" have been developed to produce multiple folds into a moving melt so as to improve mixing efficiency. The device can be fitted onto extruders to aid mixing.

Sphere in a Fluid

For a sphere in a fluid, the characteristic length-scale is the diameter of the sphere and the characteristic velocity is that of the sphere relative to the fluid some distance away from the sphere, such that the motion of the sphere does not disturb that reference parcel of fluid. The density and viscosity are those belonging to the fluid. Note that purely laminar flow only exists up to $Re = 10$ under this definition.

Under the condition of low Re, the relationship between force and speed of motion is given by Stokes' law.

Oblong Object in a Fluid

The equation for an oblong object is identical to that of a sphere, with the object being approximated as an ellipsoid and the axis of length being chosen as the characteristic length scale. Such considerations are important in natural streams, for example, where there are few perfectly spherical grains. For grains in which measurement of each axis is impractical, sieve diameters are used instead as the characteristic particle length-scale. Both approximations alter the values of the critical Reynolds number.

Fall Velocity

The particle Reynolds number is important in determining the fall velocity of a particle. When the particle Reynolds number indicates laminar flow, Stokes' law can be used to calculate its fall velocity. When the particle Reynolds number indicates turbulent flow, a turbulent drag law must be constructed to model the appropriate settling velocity.

Packed Bed

For fluid flow through a bed, of approximately spherical particles of diameter D in contact, if the "voidage" is ε and the "superficial velocity" is v_s, the Reynolds number can be defined as:

$$\mathrm{Re} \quad \frac{\rho v_s D}{\mu}.$$

or

$$\mathrm{Re} = \frac{\rho v_s D}{\mu \varepsilon}.$$

or

$$\mathrm{Re} = \frac{\rho v_s D}{\mu (1 - \varepsilon)}.$$

The choice of equation depends on the system involved: the first is successful in correlating the data for various types of packed and fludized beds, the second Reynolds number suits for the liquid-phase data, while the third was found successful in correlating the fluidized bed data, being first introduced for liquid fluidized bed system.

Laminar conditions apply up to Re = 10, fully turbulent from Re = 2000.

Stirred Vessel

In a cylindrical vessel stirred by a central rotating paddle, turbine or propeller, the characteristic dimension is the diameter of the agitator D. The velocity V is ND where N is the rotational speed. Then the Reynolds number is:

$$\mathrm{Re} = \frac{\rho N D^2}{\mu} = \frac{\rho V D}{\mu}.$$

The system is fully turbulent for values of Re above 10000.

Pipe Friction

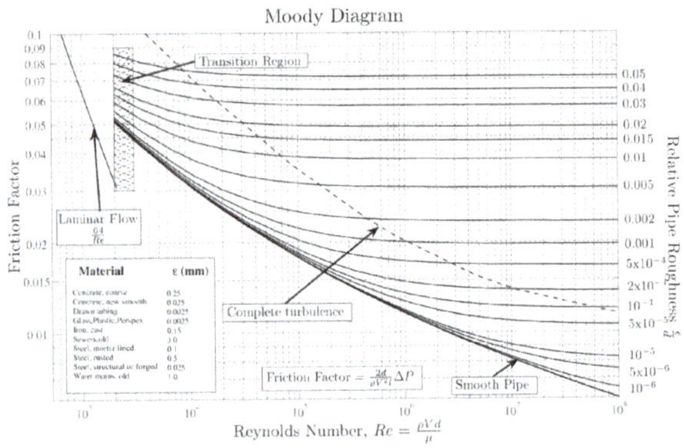

The Moody diagram, which describes the Darcy–Weisbach friction factor f as a function of the Reynolds number and relative pipe roughness.

Pressure drops seen for fully developed flow of fluids through pipes can be predicted using the Moody diagram which plots the Darcy–Weisbach friction factor f against Reynolds number Re and relative roughness ε/D. The diagram clearly shows the laminar, transition, and turbulent flow regimes as Reynolds number increases. The nature of pipe flow is strongly dependent on whether the flow is laminar or turbulent.

Similarity of Flows

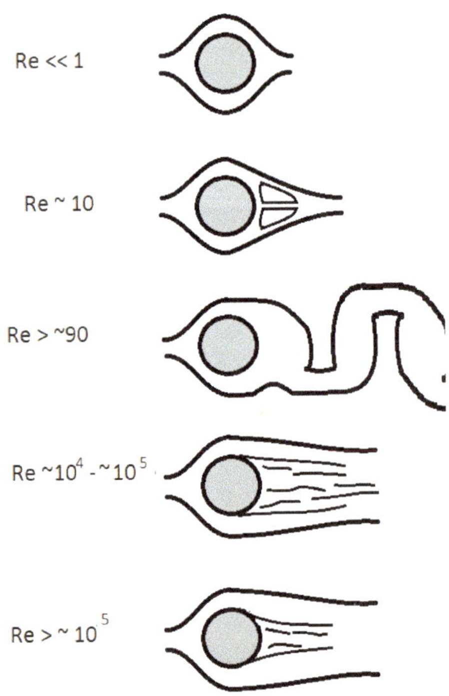

Qualitative behaviors of fluid flow over a cylinder depends to a large extent on Reynolds number; similar flow patterns often appear when the shape and Reynolds number is matched, although other parameters like surface roughness have a big effect

In order for two flows to be similar they must have the same geometry, and have equal Reynolds numbers and Euler numbers. When comparing fluid behavior at corresponding points in a model and a full-scale flow, the following holds:

$$\mathrm{Re}_{\mathrm{m}} = \mathrm{Re}$$

$$\mathrm{Eu}_{\mathrm{m}} = \mathrm{Eu}$$

Where Re_{m} is the Reynolds number for the model and Re is Reynolds number of the model; and similarly for Eu_{m} the Euler number.

The model numbers and design numbers should be in the same proportion, hence:

$$\frac{p_{\mathrm{m}}}{\rho_{\mathrm{m}} v_{\mathrm{m}}^2} = \frac{p}{\rho v^2}.$$

This allows engineers to perform experiments with reduced scale models in water channels or wind tunnels, and correlate the data to the actual flows, saving on costs during experimentation and on lab time. Note that true dynamic similitude may require matching other dimensionless numbers as well, such as the Mach number used in compressible flows, or the Froude number that governs open-channel flows. Some flows involve more dimensionless parameters than can be practically satisfied with the available apparatus and fluids, so one is forced to decide which parameters are most important. For experimental flow modeling to be useful, it requires a fair amount of experience and judgement of the engineer.

An example where the mere Reynolds number is not sufficient for similarity of flows (or even the flow regime – laminar or turbulent) are bounded flows, i.e. flows that are restricted by walls or other boundaries. A classical example of this is the Taylor–Couette flow, where the dimensionless ratio of radii of bounding cylinders is also important, and many technical applications where these distinctions play an important role. Principles of these restrictions were developed by Maurice Marie Alfred Couette and Geoffrey Ingram Taylor and developed further by Floris Takens and David Ruelle.

Typical values of Reynolds number

- Ciliate ~ 1×10^{-1}
- Smallest Fish ~ 1
- Blood flow in brain ~ 1×10^{2}
- Blood flow in aorta ~ 1×10^{3}
- Onset of turbulent flow ~ 2.3×10^{3} to 5.0×10^{4} for pipe flow to 10^{6} for boundary layers
- Typical pitch in Major League Baseball ~ 2×10^{5}
- Person swimming ~ 4×10^{6}

- Fastest Fish ~ 1×10^8
- Blue Whale ~ 4×10^8
- A large ship (RMS Queen Elizabeth 2) ~ 5×10^9

Smallest Scales of Turbulent Motion

In a turbulent flow, there is a range of scales of the time-varying fluid motion. The size of the largest scales of fluid motion (sometimes called eddies) are set by the overall geometry of the flow. For instance, in an industrial smoke stack, the largest scales of fluid motion are as big as the diameter of the stack itself. The size of the smallest scales is set by the Reynolds number. As the Reynolds number increases, smaller and smaller scales of the flow are visible. In a smoke stack, the smoke may appear to have many very small velocity perturbations or eddies, in addition to large bulky eddies. In this sense, the Reynolds number is an indicator of the range of scales in the flow. The higher the Reynolds number, the greater the range of scales. The largest eddies will always be the same size; the smallest eddies are determined by the Reynolds number.

A large Reynolds number indicates that viscous forces are not important at large scales of the flow. With a strong predominance of inertial forces over viscous forces, the largest scales of fluid motion are undamped—there is not enough viscosity to dissipate their motions. The kinetic energy must "cascade" from these large scales to progressively smaller scales until a level is reached for which the scale is small enough for viscosity to become important (that is, viscous forces become of the order of inertial ones). It is at these small scales where the dissipation of energy by viscous action finally takes place. The Reynolds number indicates at what scale this viscous dissipation occurs.

In Physiology

Poiseuille's law on blood circulation in the body is dependent on laminar flow. In turbulent flow the flow rate is proportional to the square root of the pressure gradient, as opposed to its direct proportionality to pressure gradient in laminar flow.

Using the definition of the Reynolds number we can see that a large diameter with rapid flow, where the density of the blood is high, tends towards turbulence. Rapid changes in vessel diameter may lead to turbulent flow, for instance when a narrower vessel widens to a larger one. Furthermore, a bulge of atheroma may be the cause of turbulent flow, where audible turbulence may be detected with a stethoscope.

Complex Systems

Reynolds number interpretation has been extended into the area of arbitrary complex systems. Such as financial flows, nonlinear networks, etc. In the latter case an artificial viscosity is reduced to nonlinear mechanism of energy distribution in complex network media. Reynolds number then represents a basic control parameter which expresses a balance between injected and dissipated energy flows for open boundary system. It has been shown that Reynolds critical regime separates two types of phase space motion: accelerator (attractor) and decelerator. High Reynolds number leads to a chaotic regime transition only in frame of strange attractor model.

Derivation

The Reynolds number can be obtained when one uses the nondimensional form of the incompressible Navier–Stokes equations for a newtonian fluid expressed in terms of the Lagrangian derivative:

$$\rho\frac{D\mathbf{v}}{Dt} = -\nabla p + \mu\nabla^2\mathbf{v} + \rho\mathbf{f}.$$

Each term in the above equation has the units of a "body force" (force per unit volume) with the same dimensions of a density times an acceleration. Each term is thus dependent on the exact measurements of a flow. When one renders the equation nondimensional, that is when we multiply it by a factor with inverse units of the base equation, we obtain a form which does not depend directly on the physical sizes. One possible way to obtain a nondimensional equation is to multiply the whole equation by the following factor:

$$\frac{L}{\rho V^2}$$

where:

- V is the mean velocity, v or $\bar{\mathbf{v}}$, relative to the fluid (m/s).
- L is the characteristic length (m).
- ρ is the fluid density (kg/m³).

If we now set:

$$\mathbf{v}' = \frac{\mathbf{v}}{V}, \quad p' = p\frac{1}{\rho V^2}, \quad \mathbf{f}' = \mathbf{f}\frac{L}{V^2}, \quad \frac{\partial}{\partial t'} = \frac{L}{V}\frac{\partial}{\partial t}, \quad \nabla' = L\nabla$$

we can rewrite the Navier–Stokes equation without dimensions:

$$\frac{D\mathbf{v}'}{Dt'} = -\nabla' p' + \frac{\mu}{\rho L V}\nabla'^2\mathbf{v}' + \mathbf{f}'$$

where the term $\mu/\rho LV = 1/\text{Re}$.

Finally, dropping the primes for ease of reading:

$$\frac{D\mathbf{v}}{Dt} = -\nabla p + \frac{1}{\text{Re}}\nabla^2\mathbf{v} + \mathbf{f}.$$

This is why mathematically all Newtonian, incompressible flows with the same Reynolds number are comparable. Notice also, in the above equation, as $\text{Re} \to \infty$ the viscous terms vanish. Thus, high Reynolds number flows are approximately inviscid in the free-stream.

Relationship to Other Dimensionless Parameters

There are many dimensionless numbers in fluid mechanics. The Reynolds number measures the ratio of advection and diffusion effects on structures in the velocity field, and is therefore closely related to Péclet numbers, which measure the ratio of these effects on other fields carried by the flow, for example temperature and magnetic fields. Replacement of the kinematic viscosity $\nu = \mu/\rho$ in Re by the thermal or magnetic diffusivity results in respectively the thermal Péclet number and the magnetic Reynolds number. These are therefore related to Re by products with ratios of diffusivities, namely the Prandtl number and magnetic Prandtl number.

Test Section Parameters

(i) Dynamic pressure

The local dynamic parameter $\frac{1}{2}\rho v^2$ can be related to the local Mach number and static pressure.

$$q = \frac{1}{2}\rho v^2 = \frac{1}{2}\frac{p}{RT}v^2$$
$$= \frac{1}{2}\frac{p}{\gamma RT}\gamma v^2 = \frac{1}{2}\gamma p M^2$$

(ii) Mass flow rate

$$\dot{m} = \rho v A = \rho^* a^* A^*$$

* represents choked conditions (M = 1)

$$\frac{\dot{m}}{A} = \frac{\rho A v}{A^*} * \frac{A^*}{A}$$
$$= \frac{\rho^* A^* v^*}{A^*}\frac{A^*}{A}$$

Since, $\rho^* = \frac{p^*}{RT^*}$ and $v^* = a^* = \sqrt{\gamma RT^*}$

One has

$$\frac{\dot{m}}{A} = \frac{p^*}{RT^*}\sqrt{\gamma RT^*}\frac{A^*}{A}$$
$$= \sqrt{\frac{\gamma}{R}}\frac{p^*}{\sqrt{T^*}}\frac{A^*}{A}$$

Using equation for $\frac{T^*}{T_0}$ *and* $\frac{p^*}{p_0}$

and writing in terms of stagnation conditions

$$\frac{T_0}{T} = 1 + \frac{\gamma - 1}{2} M^2$$

$$\frac{T_0}{T^*} = 1 + \frac{\gamma - 1}{2} = \frac{2 + \gamma - 1}{2} = \frac{\gamma + 1}{2}$$

When M = 1,

$$\frac{p_0}{p^*} = \left(\frac{\gamma + 1}{2} \right)^{\frac{\gamma}{\gamma - 1}}$$

$$\frac{\dot{m}}{A} = \sqrt{\frac{\gamma}{R}} \frac{p_0}{\sqrt{T_0}} \frac{\left(\frac{2}{\gamma + 1} \right)^{\frac{\gamma}{\gamma - 1}}}{\left(\frac{2}{\gamma - 1} \right)^{1/2}} \frac{A^*}{A}$$

$$\frac{\dot{m}}{A} = \sqrt{\frac{\gamma}{R}} \frac{p_0}{\sqrt{T_0}} \left(\frac{2}{\gamma + 1} \right)^{\frac{\gamma + 1}{2(\gamma - 1)}} \frac{A^*}{A}$$

Mass flow density is expressed as a function of stagnation conditions and the area ratio. For isentropic flow, A^* is a constant and $\frac{A}{A^{\cdot}}$ is a unique function of local Mach number.

$$\frac{A}{A^{\cdot}} = \frac{1}{M_1} \left[\frac{1 + \frac{\gamma - 1}{2} M^2}{\frac{\gamma + 1}{2}} \right]^{\gamma + 1(\gamma - 1)}$$

$$\frac{\dot{m}}{A} = \sqrt{\frac{\gamma}{R}} \frac{p_0}{\sqrt{T_0}} M \left[1 + \frac{\gamma - 1}{2} M^2 \right]^{\frac{\gamma + 1}{-2(\gamma - 1)}}$$

(iii) Test section velocity

V = a M

$$= \left[\frac{\gamma R T_0 M^2}{1 + \frac{\gamma - 1}{2} M^2} \right]^{\frac{1}{2}}$$

(iv) Maximum velocity

$$C_p T + \frac{v^2}{2} = C_p T_0$$

v is maximum when $c_p T = 0$

$$C_p - C_v = R$$

$$C_p / C_v = \gamma$$

$$C_p - \frac{C_p}{\gamma} = R$$

$$\frac{C_p \gamma - C_p}{\gamma} = R$$

$$\frac{C_p(\gamma - 1)}{\gamma} = R$$

$$C_p = \frac{\gamma R}{\gamma - 1}$$

$$v_{max}^2 = 2C_p T_0$$

$$v_{max} = \left[\frac{2\gamma R}{\gamma - 1} T_0\right]^{\frac{1}{2}}$$

The test section flow velocity v for a given stagnation temperature T_O approaches the maximum value v_{max} at relatively low supersonic Mach numbers.

For example,

In the case of $T_O = 300K$

$R = 287 J/kgK$

and a test section Mach number of 5.0, the ratio of v/v_{max} can be calculated to see that it is equal to 0.913. This means at ordinary stagnation temperatures, the velocity in the test section reaches 91% of the maximum possible velocity corresponding to the total energy of the fluid. The stagnation temperature T_O rather than the Mach number which is important to attain high velocities.

Free Stream Reynolds Number (Re)

$$\mathrm{Re} = \frac{\rho v L}{\mu}$$

Experimental observation is that μ is independent of pressure in the range of 0.001 to 20 atmospheres.

$$\left(\frac{\mu}{\mu_0}\right) = \left(\frac{T}{T_0}\right)^n \quad \text{for air n = 0.768}$$

$$\frac{\mu}{\mu_0} = \frac{T_0 + C}{T + C}\left(\frac{T}{T_0}\right)$$

C=130

$$\frac{\rho}{\rho_0} = \left(\frac{T}{T_0}\right)^{\frac{1}{\gamma-1}}$$

If this relation is assumed then the free stream Re can be expressed as a function of M_1, the T.S Mach no and of the stagnation parameters. Reynolds no per unit length $\frac{Re}{L} = \frac{\rho v}{\mu} \rightarrow \rho, v, \mu$ are expressed in stagnation quantities as given below.

$$\frac{\rho}{\rho_0} = \left[1 + \frac{\gamma - 1}{2} M^2\right]^{-\frac{1}{\gamma-1}}$$

$$\rho = \rho\left[1 + \frac{\gamma - 1}{2} M^2\right]^{-\frac{1}{\gamma-1}}$$

$$\mu = \mu_0 \left(\frac{T}{T_0}\right)^n = \mu_0 \left(\frac{1}{1 + \frac{\gamma - 1}{2} M^2}\right)^n$$

$$M = \frac{v}{a}$$

$$v = Ma$$

$$= Ma\left(\frac{1}{1 + \frac{\gamma - 1}{2} M^2}\right)^{\frac{1}{2}}$$

$$\frac{a^2}{a_0^2} = \frac{\gamma RT}{\gamma RT_0}$$

$$\frac{Re}{L} = \frac{\rho_0\left[1 + \frac{\gamma - 1}{2} M^2\right]^{-\frac{1}{\gamma-1}}}{\mu_0\left[\frac{1}{1 + \frac{\gamma - 1}{2} M^2}\right]} * M\, a_0 \left(\frac{1}{1 + \frac{\gamma - 1}{2} M^2}\right)^{\frac{1}{2}}$$

n= 0.768 for air

air Simplifying

$$\frac{\text{Re}}{L} = \frac{\rho_0}{\mu_0} a_0 M \left[1 + \frac{\gamma - 1}{2} M^2\right]^{0.268 - \frac{1}{\gamma - 1}}$$

Both a_0 and μ_0 are functions of stagnation temperature. Both increase with temperature. Hence, appreciable changes in free stream Re/unit length for a given M can be obtained only by varying stagnation density.

Components of Supersonic Wind Tunnels

Air Storage Tanks

Size of the storage will be dependent on the mass flows required and the frequency of runs. Pressure storage tanks are available on the shelf basis – They are mounted horizontally or vertically. Tanks are painted black to absorb heat. They are provided with safety disk or pressure relief valve. As air is drawn from the storage, polytropic expansion takes place within the tank. This results in drop of reservoir temperature which is very bothersome. Fall of stagnation temperature causes resultant change in the stream temperature for a given Mach number. Change in temperature results in the change of viscosity which in turn affects the boundary layer thickness. Changes in Reynolds number and Mach number during a run are thus consequential to the fall in reservoir temperature.

To maintain constancy of stagnation temperature, it is a practice to stack the reservoir volume with empty metallic cans. They serve as heat storing matrix during compression and release heat during the expansion process. Another way to maintain the constant stagnation temperature is by providing heater units in the reservoir.

Settling Chamber /wide Angle Diffusers

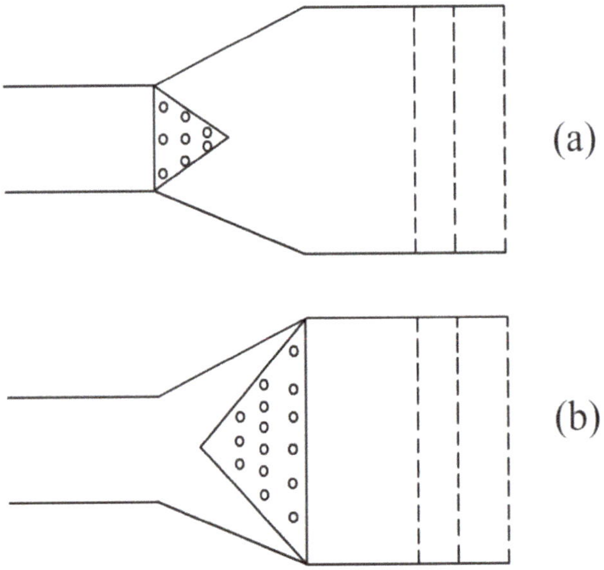

Wide angle diffuser

Wide angle diffusers lead the flow to the settling chamber. Arrangements for leading the flow to the settling chamber may be by one of the methods shown in Figure (a, b or c).

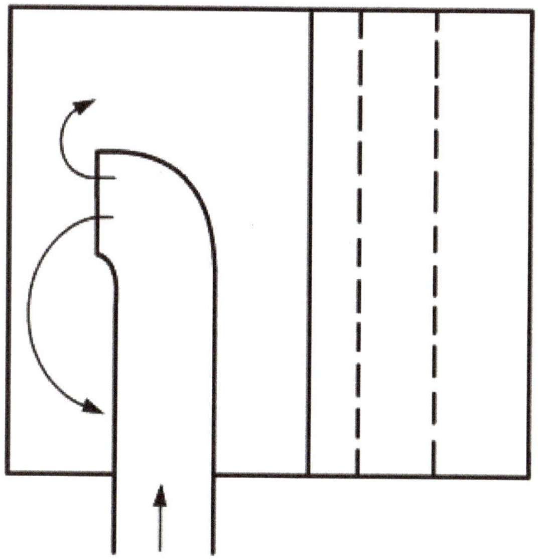

(c) Reverse entry into the settling chamber

Uniformity of flow in the test section is improved if a large area ratio contraction is provided.

Convergent-divergent (c-d) Nozzle

The c-d nozzle forms the heart of the supersonic wind tunnel .For generating supersonic flow in the test section, it is essential that there is a c-d nozzle in the tunnel circuit before the test section. The area ratio of the c-d nozzle (A_{exit}/A_{throat}) uniquely decides the Mach number.

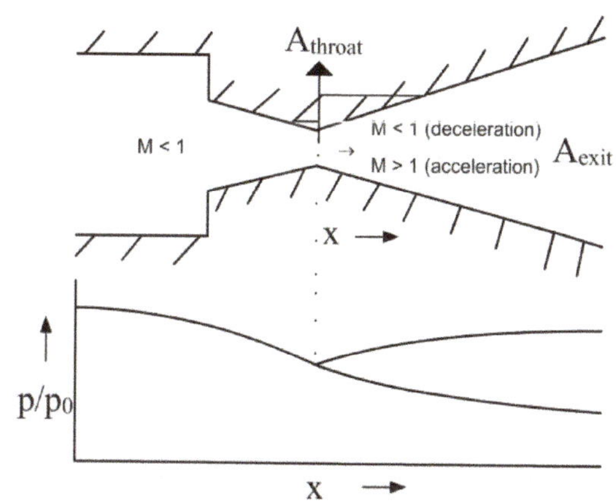

(a) Convergent divergent nozzle and the pressure profile

When the tunnel operation starts, the flow is initiated in the nozzle as subsonic and reaches the sonic Mach number at the throat when sufficient mass flow is allowed. This is called the choked condition of the nozzle. Under this condition, maximum mass flow rate for the given stagnation conditions takes place through the nozzle. The ratio between the upstream stagnation pressure (P_o) and the downstream back pressure (P_b) corresponding to the first time choking is called the first critical pressure ratio of the nozzle. At this pressure ratio, the flow in the divergent part of the nozzle is subsonic. The exit Mach number will be the subsonic value corresponding to the A/A* in the Fig. (b). In this context A is the exit area and A* the throat area of the choked nozzle. As the value of P_o/P_b is progressively increased, the flow in the divergent part of the nozzle accelerates to be supersonic but shocks are formed in the divergent part until a pressure ratio corresponding to the supersonic Mach number of the nozzle is reached. The pressure ratio corresponding to this Mach number is the third critical pressure ratio of the nozzle. Between the first and third critical pressure ratios shocks of varying strengths take place in the nozzle and outside of it as there is no other isentropic solution between the 1st and 3rd critical pressure ratios. The pressure ratio corresponding to the occurrence of a shock at the exit plane of the nozzle is the second critical pressure ratio of the nozzle.

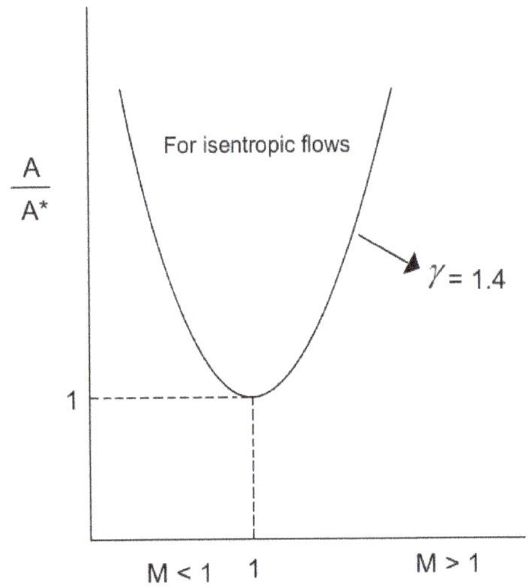

(b) A/A* vs. Mach numbers

All the shocks generated at the different pressure ratios inside the nozzle will make the post shock Mach number subsonic and the subsonic nozzle exit pressure will be made equal to the ambient pressure in the remaining part of the diffusing divergent channel. Between the second and third critical pressure ratios, oblique shocks of varying strengths depending on the pressure ratio will be formed emanating from the nozzle lip. The physical purpose of these oblique shocks is equalization of pressures between the exit plane and the ambient. If the pressure ratio is increased beyond that corresponding to the third critical pressure ratio, expansion fans will be formed at the lip of the nozzle.

Diffuser- the Necessity of Providing a Diffuser

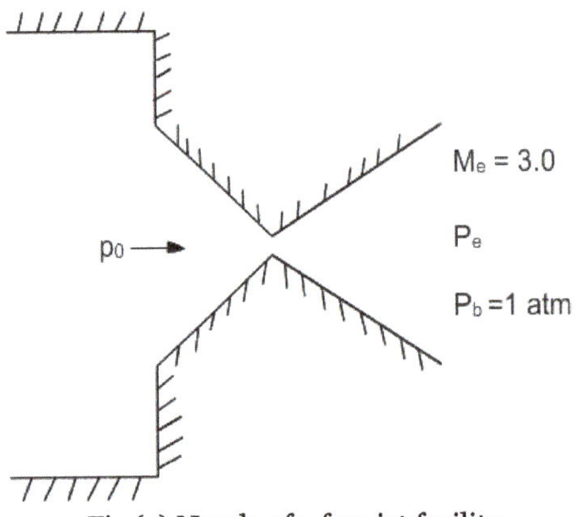

Fig (a) Nozzle of a free jet facility

Take the case of a free jet facility as in the given figures making use of a c-d nozzle of Mach number 3.0 exiting to the ambient conditions at one atmosphere pressure. In order to avoid shocks and expansion waves at the exit of the nozzle, p_e must be p_b.

$$\left(\frac{p_o}{p_e}\right)_{M=3} = 36.7$$

Pressure ratio required will be $\frac{p_o}{pe} = 36.7$ for a wave free exit flow from the nozzle.

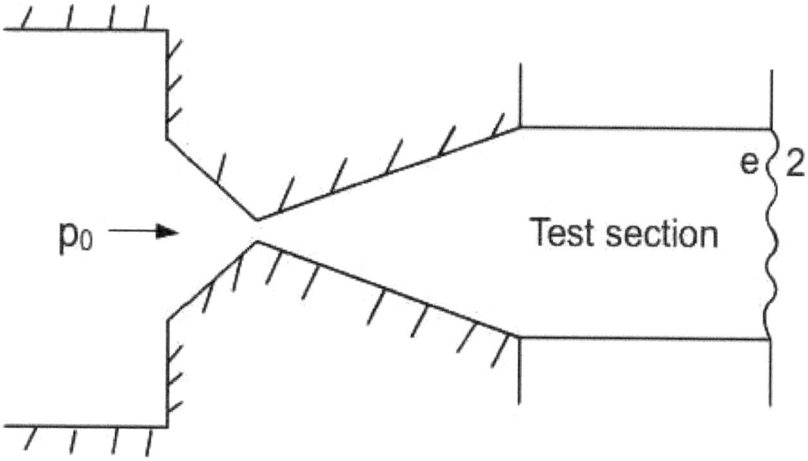

Fig (b) Free jet nozzle with a test section

In the Figure (b) a constant area section is added to the nozzle exit. The duct similar to the test section (T.S) of a wind tunnel attached to the nozzle exhausts to atmosphere. p_e corresponds to static pressure at the exit plane of the nozzle before the shock. Static pressure after the shock (p_2) is equal to ambient pressure.

$$\frac{p_o}{p_2} = \left(\frac{p_o}{p_e}\right) \times \left(\frac{p_e}{p_2}\right) = 36.7 \times \left(\frac{1}{10.33}\right) = 3.55$$

In the equation above, p_e/p_2 represents the shock pressure ratio at M=3.0.

In the third case, as in the figure given below divergent channel is provided after the constant area duct and the shock stands at the end of the constant area duct.

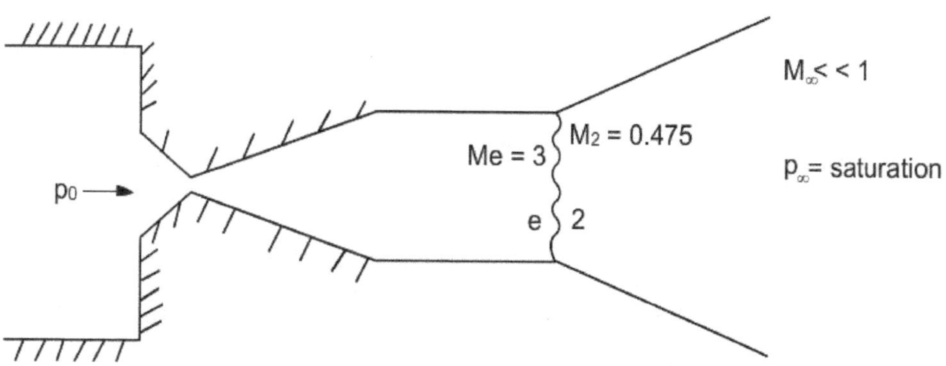

Free jet facility with test section and a diffuser

When $M << 1$, $p_o = p_\infty$

$$\frac{p_o}{p_\infty} = \left(\frac{p_0}{p_e}\right)\left(\frac{p_e}{p_2}\right)\left(\frac{p_2}{p_{o2} = p_\infty}\right)_{M=0.475} = 36.7 \times \frac{1}{10.33} \times 0.856 = 3.04$$

The three cases described above make it clear that provision of a diffuser of suitable design is required for reducing the pressure ratio required for the operation of the wind tunnel. The power required to run the wind tunnel increases with the pressure ratio. In supersonic wind tunnels, most commonly used diffuser is of convergent divergent type (also called the second throat diffuser).

Power Required for the Operation of Supersonic Wind Tunnel

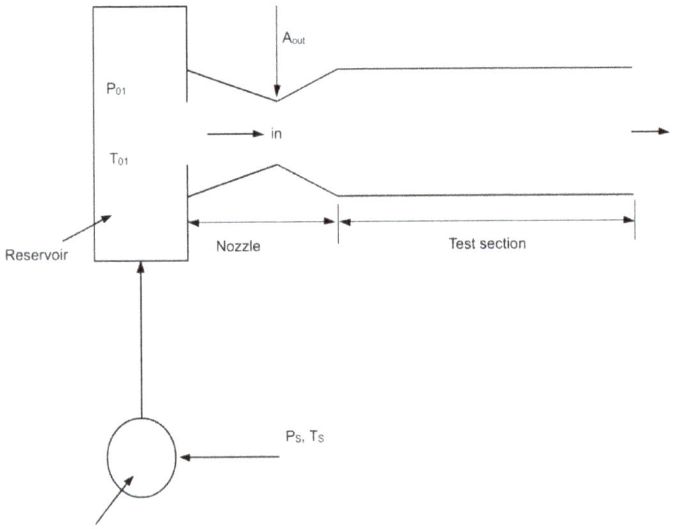

Free jet type wind tunnel with an attached test section

Refer to the figure where a free jet type wind tunnel is shown. Let the supersonic tunnel is specified by the Mach number (M) in the test section and test section (A) area .The throat area is specified as A_{NT}. The flow parameters in the test section are denoted as p, T, A etc.. Once M is specified, the area ratio A_T/A_{NT} and ratios of pressure and temperature p/p_{01} and T/T_{01} are all known from the isentropic equations. If the compressor is idealized (isentropic) the suction and reservoir conditions are related.

$$\frac{T_S}{T_{01}} = \left(\frac{p_S}{p_{01}}\right)^{\gamma-1/\gamma}$$

Where p_S and T_s represent pressure and temperature at the compressor suction. If the nozzle is choked, the mass flow rate can be written as,

$$\dot{m} = A_{NT}\sqrt{\frac{\gamma}{R}}\left(\frac{2}{\gamma+1}\right)^{(\gamma+1)/2(\gamma-1)} \frac{p_{01}}{\sqrt{T_0}}$$

The power required to operate the compressor may be found as follows:

If the compressor is isentropic, the work per unit time may be found from the enthalpy-difference across the compressor.

$$\dot{W} = -\dot{m}(h_{01} 0 h_s)$$

$$= -\dot{m}\frac{\gamma R}{\gamma - 1}(T_{01} - T_S) \qquad \left[\dot{W} = work\,/\,unit\ time\ in\ W\ or\ \frac{Nm}{S}\right]$$

$$= \frac{\dot{m}\gamma R T_s}{\gamma - 1}\left(\frac{T_{01}}{T_S} - 1\right)$$

$$= -\dot{m}\gamma\frac{RT_S}{\gamma - 1}\left[\left(\frac{p_{01}}{p_S}\right)^{\frac{\gamma-1}{\gamma}} - 1\right]$$

$$\dot{W} = -\dot{m}\frac{\gamma RT_S}{\gamma - 1}\left[(r_p)^{(\gamma-1)/\gamma} - 1\right]$$

The pressure ratio $\frac{p_{01}}{p_s}$ is known as the operating pressure ratio and is denoted by r_p

The very large power required for the operation of a supersonic wind tunnel is attributed to the large operating pressure ratio. If the wind tunnel is equipped with a suitably designed diffuser and a closed circuit arrangement as shown in the next section, the stagnation pressure of the diffused high velocity air can be made use of by the compressor and the effective pressure ratio can be reduced.

Closed Circuit Supersonic Wind Tunnel

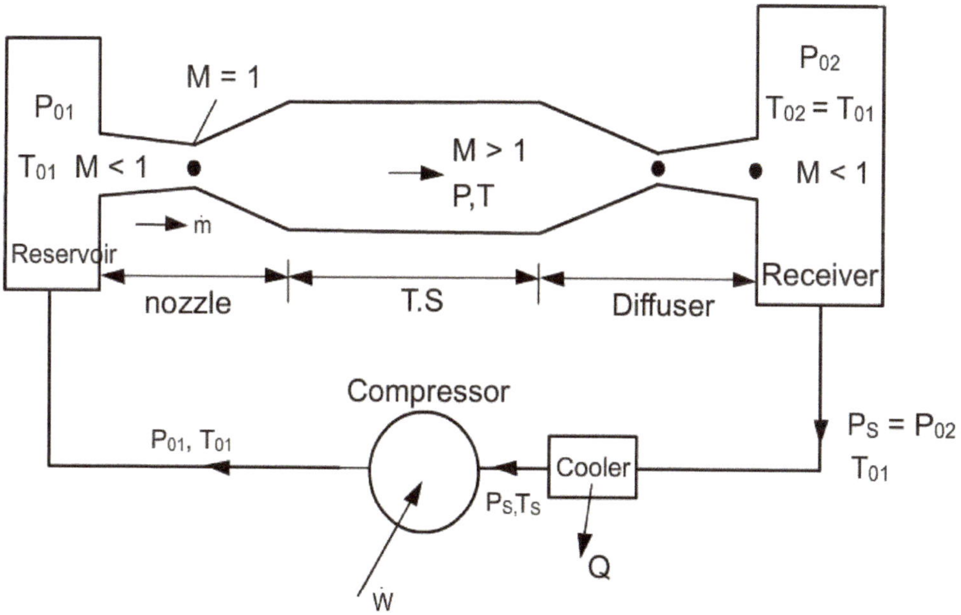

Continuous type supersonic wind tunnel

A convergent-divergent (c-d) diffuser is provided as shown in the figure above. Only if the flow is isentropic, the stagnation pressure regained in the receiver following the diffuser (p_{02}) is equal to that of the flow entering the nozzle (p_{01}).

In that case, $p_{02} = p_{01}$

Then, $r_p = 1$, so that compressor does no work. In practical cases, because of entropy changes $p_{02} < p_{01}$ and $r_p > 1$. If the entropy change is confined to the region between the two throats, the diffuser throat area A_{DT} must be larger than the nozzle throat area A_{NT}. Diffuser throat area must be large enough to accommodate the stagnation pressure loss of the strongest shock. A cooler is included prior to the compressor because compressor work is proportional to the intake temperature.

The practical operation of the closed circuit wind tunnel may be explained as follows: As the tunnel is started, flow through it begins as subsonic and as the pressure ratio is increased the nozzle is choked. Further increase in the pressure ratio causes shock to be formed in the divergent section. At a pressure ratio corresponding to second critical pressure ratio, shock is formed at the exit plane of the nozzle which is same as the entry section to the test section. The formation of shocks during the starting process gives rise to fall in stagnation pressure. The total pressure after the shock is designated as p_{02}. This necessitates that the diffuser throat is designed larger as decided by the ratio p_{02}/p_{01}. The ratio of diffuser throat area to the nozzle throat is in the inverse ratio of total pressures given above.

$$\frac{A_{NT}}{A_{DT}} = \frac{p_{02}}{p_{01}} < 1$$

The ratio of areas for different test section Mach numbers calculated based on normal shock losses is given in the figure below. This makes sure that the starting shock passes through the diffuser throat. The diffuser throat area calculated as above does not take in to account the non- isentropy of frictional flows and only the shock losses are considered.

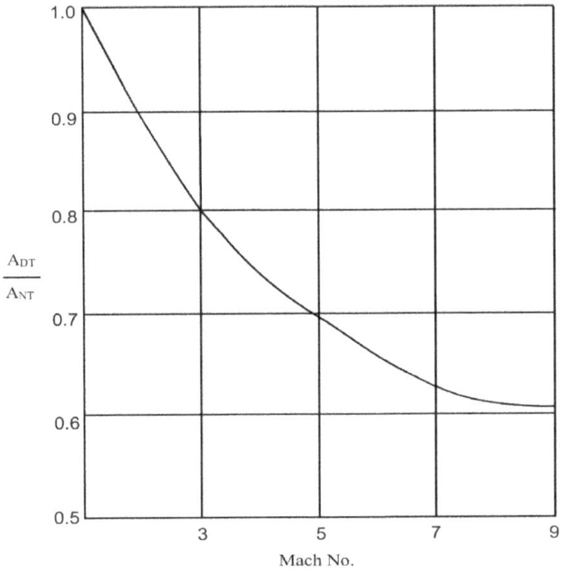

Ratio of diffuser throat and nozzle throat for different test section Mach numbers

Assuming a frictionless operation, the shock may assume any section in the constant area test section. But, the effect of friction is to make the shock unstable in the constant area duct. The shock that is generated during starting of the tunnel does not stay at the nozzle exit (entry to test section) but is moved downstream by the effect of friction.

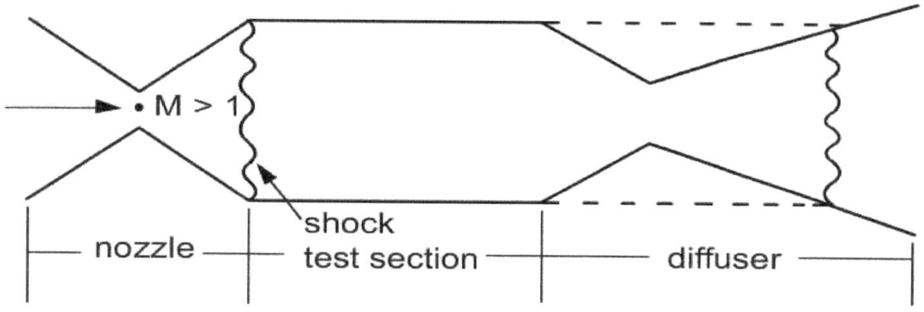

Shock movement from nozzle exit to diffuser

The starting pressure ratio minimum required to cater to the shock at the test section Mach number is corresponding to that for locating the shock at the nozzle exit(worst shock).

$$r_{ps} = \left(\frac{p_{01}}{p_{02}}\right)_{worst\ shock}$$

As the starting pressure ratio is maintained, the starting shock which moves down stream can be stable only at an area equal to that of the test section. In the convergent part of the diffuser the Mach number will be less than that in the test section. With the value of starting pressure ratio which produced the shock at the test section Mach number being maintained, shock losses at that Mach number is being catered to. Hence the starting shock stabilizes only in the diverging part of the diffuser at a section where there is equal area and Mach number as the test section. The starting shock crossing the diffuser throat and remaining in its divergent part is called the 'swallowing of the starting shock'.

It has to be remembered that as the diffuser throat is larger than the nozzle throat, the Mach number there will be more than one but less than that in the test section. After the brief duration of starting, the pressure ratio may be decreased and the shock may be brought to the diffuser throat. Therefore the higher pressure ratio is required only during the starting when a shock at the test section Mach number is necessarily to be catered to and thereafter the pressure ratio can be that corresponding to shock at the diffuser throat. The smallest pressure ratio at which the tunnel may be continuously operated (rPo) after its starting is that corresponding to the stagnation pressure loss of the weakest starting shock which can be made to occur at the diffuser throat.

$$r_{PO} = \left(\frac{p_{01}}{p_{02}}\right)_{Shock\ at M_{DT}}$$

In summary, the pressure ratio for starting the wind tunnel is corresponding to the normal shock losses at the test section Mach number and that for operation is that corresponding to normal shock losses at the diffuser throat. Hence the power required can be considerably reduced by incorporating the well designed diffuser and by judicious control of the two pressure ratios during 'starting 'and 'operating' of the wind tunnel.

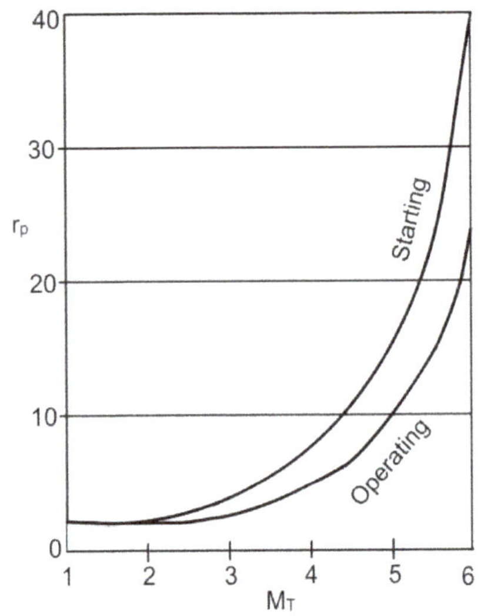

Pressure ratios for starting and operating the wind tunnel of different Mach numbers

Actual Flow in the Supersonic Wind Tunnel

The boundary layer (B.L) thickness and the total loss of momentum increase with increasing distance from the 1st throat. The growth of boundary layer thickness with distance from first throat is predictable and can be accounted for in the nozzle design. In the steady state operation, viscous effects between the throat and test section are not of much importance.

During the transient process in which tunnel is started, viscous effects are much important. So, important are these effects that pressure ratios required to start high Mach number tunnels are atleast 100% greater than the normal shock pressure ratio $\frac{p_{01}}{p_{02}}$ i.e. viscous losses are almost equal to normal shock losses.

B.L. is stable when the pressure is decreasing in the direction of its growth. When the pressure is increasing in the direction of flow, it has a tendency to separate. As normal shock passes through the nozzle, it imposes a severe unfavorable pressure gradient which can cause separation. If B.L separates, it disturbs the flow over a large portion of nozzle. If B.L. does not separate the high pressure gain in the downstream of shock will tend to flow to low pressure B.L. and the flow in the duct will be altered over a significant length of the nozzle. In the diffuser viscous effects are predominant during starting and steady state operation of the wind tunnel. Unfavorable pressure gradient exists always. An oblique shock from the convergence creates additional pressure gradients when they strike the opposite wall.

Starting a Tunnel with a Model in the Test Section

It is explained earlier that

$$\frac{A_{DT}}{A_{NT}} = \frac{p_{01}}{p_{02}}$$

which implies that losses in total head resulting from shocks necessitate a larger diffuser throat. Hence, losses due to shocks on the model must also be provided for. So, for starting a tunnel with a model, a second throat larger than that for a clear tunnel is needed.

Sizing the Wind Tunnel Model

The theoretical unobstructed cross section area of the test section at the model required for starting is the same as the second throat area.

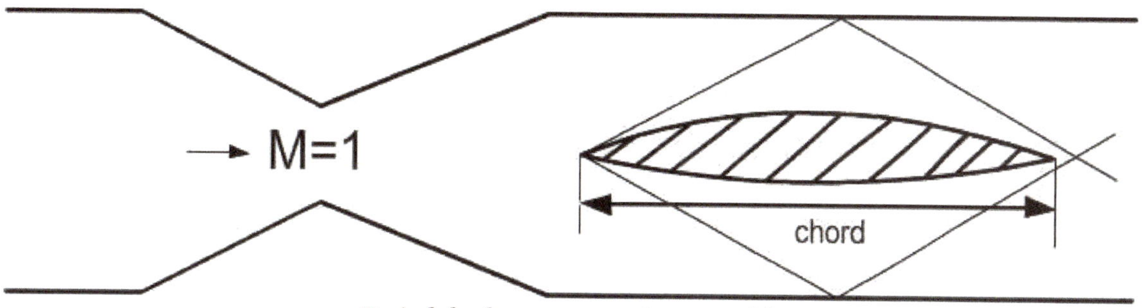

Typical shock wave pattern from a model

In choosing the dimensions of the model, reflection of shocks should be also considered. The oblique shocks formed as shown in the figure at the leading edge of the model get reflected from the wind tunnel wall. It has to be remembered that shock reflection is not specular which means that the angle of incidence of the shock at the test section wall is not same as that for reflection. The chord length of the model is so chosen that the reflected shocks do not interfere with the model.

Two Problems in the Operation of Supersonic Wind Tunnels

(i) Condensation

The amount of moisture that can be held by a unit volume of air increases with increasing temperature. When the air isentropically expands to higher Mach numbers in the test section, the temperature falls. It may become super cooled. Moisture will then condense.

Factors affecting condensation

- Amount of moisture in the stream
- Static temperature of the stream
- Static pressure of the stream
- Time during which the stream is at low temperature

Effects of condensation

Condensation results in changes of local Mach number and other flow properties due to latent heat addition. The extent of changes depends on how much heat is released through condensation and may be evaluated using the two equations given below:

$$\frac{dM^2}{M^2} = \frac{1+\gamma M^2}{\left(\gamma - M^2\right)}\left[\frac{dQ}{H} - \frac{dA}{A}\right]$$

$$\frac{dp}{p} = -\frac{\gamma M^2}{1-M^2}\left[\frac{dQ}{H} - \frac{dA}{A}\right]$$

The notations used in the equations are:

dQ = heat added through condensation

H = enthalpy per unit mass

A = duct area

When M > 1, Mach no decreases and pressure increases. When M < 1, Mach no increases and pressure decreases.

Drying the working fluid is the best way to avoid condensation. Increasing the temperature by providing stagnation heaters is another solution.

(ii) Liquefaction

In a manner parallel to condensation, the components of air liquefy when proper temperature and pressure conditions are met. Liquefaction troubles might start around M=4 if high pressure air is expanded from room temperature.

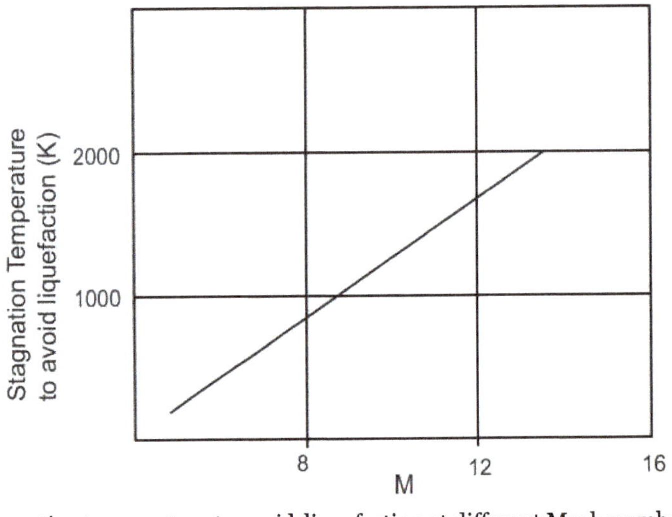

Stagnation temperature to avoid liquefaction at different Mach numbers

Figure given above shows the stagnation temperature required to avoid liquefaction at different Mach numbers. It can be seen that corresponding to a Mach number above 12 the temperature required will be about 2000K.

References

- Eckardt, Dietrich: "The 1x1 m hypersonic wind tunnel Kochel/Tullahoma 1940-1960", CEAS Space Journal, March 2015, Vol. 7, Issue 1, pp. 23-3

- Uwe Ehrenstein; Christophe Eloy. "Skin friction on a moving wall and its implications for swimming animals" (PDF). Retrieved 11 February 2014
- Houghton, E.L. and Carpenter, P.W. (1993), Aerodynamics for Engineering Students, Butterworth and Heinemann, Oxford UK. ISBN 0-340-54847-9
- Tansley, Claire E.; Marshall, David P. (2001). "Flow past a Cylinder on a Plane, with Application to Gulf Stream Separation and the Antarctic Circumpolar Current" (PDF). Journal of Physical Oceanography. 31 (11): 3274–3283. Bibcode:2001JPO....31.3274T. doi:10.1175/1520-0485(2001)031<3274:FPACOA>2.0.CO;2
- P. B. S. Lissaman (1983). "Low-Reynolds-Number Airfoils,". Annu. Rev. Fluid Mech. (15): 223–39. Bibcode:1983AnRFM..15..223L. doi:10.1146/annurev.fl.15.010183.001255
- ISBN 978-0-07-148781-8Essential Principles of Physics, P.M. Whelan, M.J. Hodgeson, 2nd Edition, 1978, John Murray, ISBN 0-7195-3382-1
- Patel, V. C.; Rodi, W.; Scheuerer, G. (1985). "Turbulence Models for Near-Wall and Low Reynolds Number Flows—A Review". AIAA Journal. 23 (9): 1308–1319. Bibcode:1985AIAAJ..23.1308P. doi:10.2514/3.9086
- Dwivedi, P. N. (1977). "Particle-fluid mass transfer in fixed and fluidized beds". Industrial & Engineering Chemistry Process Design and Development. 16 (2): 157–165. doi:10.1021/i260062a001
- Fox, R. W.; McDonald, A. T.; Pritchard, Phillip J. (2004). Introduction to Fluid Mechanics (6th ed.). Hoboken: John Wiley and Sons. p. 348. ISBN 0-471-20231-2
- Reynolds, Osborne (1883). "An experimental investigation of the circumstances which determine whether the motion of water shall be direct or sinuous, and of the law of resistance in parallel channels". Philosophical Transactions of the Royal Society. 174 (0): 935–982. JSTOR 109431. doi:10.1098/rstl.1883.0029
- Sinnott, R. K. Coulson & Richardson's Chemical Engineering, Volume 6: Chemical Engineering Design (4th ed.). Butterworth-Heinemann. p. 73. ISBN 0-7506-6538-6
- Rott, N. (1990). "Note on the history of the Reynolds number". Annual Review of Fluid Mechanics. 22 (1): 1–11. Bibcode:1990AnRFM..22....1R. doi:10.1146/annurev.fl.22.010190.000245
- Dusenbery, David B. (2009). Living at Micro Scale. Cambridge, Mass.: Harvard University Press. p. 136. ISBN 9780674031166

2

An Integrated Study of Shock Tube

The instrument which is used to replicate an explosion in order to understand its effects is known as a shock tube. It is usually rectangular in shape and is made up of metal. Shock tubes are used to create and direct blast waves towards a particular object to understand the intensity of the explosion. The major components of shock tube are discussed in this section.

Shock Tube

The shock tube is a device in which a plane propagating shock is produced in a long tube by the abrupt rupturing of the diaphragm separating one section of the tube at higher pressure than that in the other section. This device, developed after a series of studies and experiments in mid 1800s started with the realization that waves from explosions travel at a velocity faster than the sound waves. In 1948 Sir George Stokes discussed the instability of finite amplitude sound waves. Paul Vieille (1899) measured the speed of the pressure pulse generated in a glass tube in which a thin diaphragm across a pressure ratio of 27atm was burst. This pioneering work on bursting diaphragm shock tube led others to follow similar experiments with different aims. Payman, Shepherd and other colleagues of Mines Research Board did a series of investigations on the ignition of explosive gas mixtures by shock waves. The first research paper making use of shock tube was published by the Royal Society and authored by Payman and Shepherd (1946). During the post 2nd world war period, led by many Universities around the globe the shock tube was developed as a tool for aeronautical research.

Shock tube test apparatus at the University of Ottawa, Canada.

The shock tube is an instrument used to replicate and direct blast waves at a sensor or a model in order to simulate actual explosions and their effects, usually on a smaller scale. Shock tubes (and related impulse facilities such as shock tunnels, expansion tubes, and expansion tunnels) can also be used to study aerodynamic flow under a wide range of temperatures and pressures that are difficult to obtain in other types of testing facilities. Shock tubes are also used to investigate compressible flow phenomena and gas phase combustion reactions. More recently, shock tubes have been used in biomedical research to study how biological specimens are affected by blast waves.

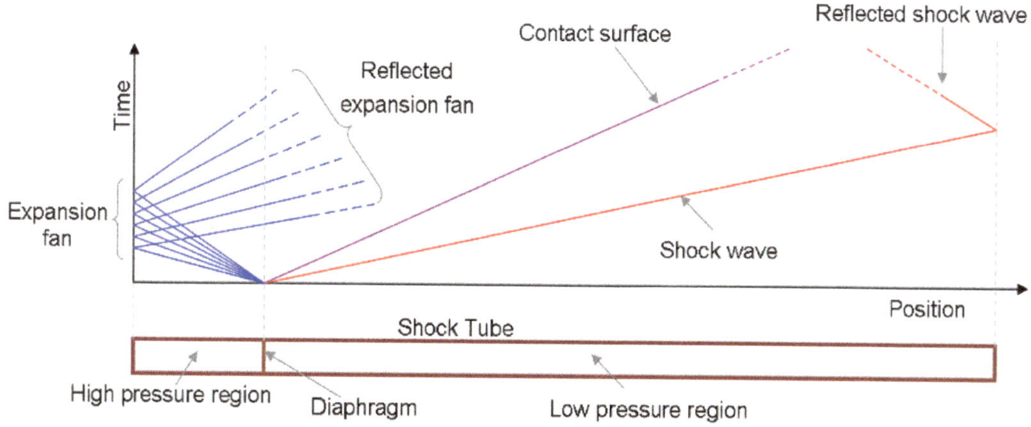

An idealized shock tube. The plot shows different waves which are formed in the tube once the diaphragm is ruptured.

A shock wave inside a shock tube may be generated by a small explosion (blast-driven) or by the buildup of high pressures which cause diaphragm(s) to burst and a shock wave to propagate down the shock tube (compressed-gas driven).

History

An early study of compression driven shock tubes was published in 1899 by French scientist Paul Vieille, though the apparatus was not called a shock tube until the 1940s. In the 1940s, interest revived and shock tubes were increasingly used to study the flow of fast moving gases over objects, the chemistry and physical dynamics of gas phase combustion reactions. In 1966, Duff and Blackwell described a type of shock tube driven by high explosives. These ranged in diameter from 0.6 to 2 m and in length from 3 m to 15 m. The tubes themselves were constructed of low-cost materials and produced shock waves with peak dynamic pressures of 7 MPa to 200 MPa and durations of a few hundred microseconds to several milliseconds.

Both compression-driven and blast-driven shock tubes are currently used for scientific as well as military applications. Compressed-gas driven shock tubes are more easily obtained and maintained in laboratory conditions; however, the shape of the pressure wave is different from a blast wave in some important respects and may not be suitable for some applications. Blast-driven shock tubes generate pressure waves that are more realistic to free-field blast waves. However, they require facilities and expert personnel for handling high explosives. Also, in addition to the initial pressure wave, a jet effect caused by the expansion of compressed gases (compression-driven) or production of rapidly expanding gases (blast-driven) follows and may transfer momentum to a sample after the blast wave has passed. More recently, laboratory scale shock tubes driven by fuel-air mixtures have been developed that produce realistic blast waves and can be operated in more ordinary laboratory facilities. Because the molar volume of gas is much less, the jet effect is a fraction of that for compressed-gas driven shock tubes. To date, the smaller size and lower peak pressures generated by these shock tubes make them most useful for preliminary, nondestructive testing of materials, validation of measurement equipment such as high speed pressure transducers, and for biomedical research as well as military applications.

Operation

A simple shock tube is a tube, rectangular or circular in cross-section, usually constructed of metal, in which a gas at low pressure and a gas at high pressure are separated using some form of diaphragm. See, for instance, texts by Soloukhin, Gaydon and Hurle, and Bradley. The diaphragm suddenly bursts open under predetermined conditions to produce a wave propagating through the low pressure section. The shock that eventually forms increases the temperature and pressure of the test gas and induces a flow in the direction of the shock wave. Observations can be made in the flow behind the incident front or take advantage of the longer testing times and vastly enhanced pressures and temperatures behind the reflected wave.

The low-pressure gas, referred to as the driven gas, is subjected to the shock wave. The high pressure gas is known as the driver gas. The corresponding sections of the tube are likewise called the driver and driven sections. The driver gas is usually chosen to have a low molecular weight, (e.g., helium or hydrogen) for safety reasons, with high speed of sound, but may be slightly diluted to 'tailor' interface conditions across the shock. To obtain the strongest shocks the pressure of the driven gas is well below atmospheric pressure (a partial vacuum is induced in the driven section before detonation).

The test begins with the bursting of the diaphragm. Several methods are commonly used to burst the diaphragm.

- A mechanically-driven plunger is sometimes used to pierce it or an explosive charge may be used to burst it.
- Another method is to use diaphragms of plastic or metals to define specific bursting pressures. Plastics are used for the lowest burst pressures, aluminum and copper for somewhat higher levels and mild steel and stainless steel for the highest burst pressures. These diaphragms are frequently scored in a cross-shaped pattern to a calibrated depth to ensure that they rupture evenly, contouring the petals so that the full section of the tube remains open during the test time.
- Yet another method of rupturing the diaphragm utilizes a mixture of combustible gases, with an initiator designed to produce a detonation within it, producing a sudden and sharp increase in what may or may not be a pressurized driver. This blast wave increases the temperature and pressure of the driven gas and induces a flow in the direction of the shock wave but at lower velocity than the lead wave.

The bursting diaphragm produces a series of pressure waves, each increasing the speed of sound behind them, so that they compress into a shock propagating through the driven gas. This shock wave increases the temperature and pressure of the driven gas and induces a flow in the direction of the shock wave but at lower velocity than the lead wave. Simultaneously, a rarefaction wave, often referred to as the Prandtl-Meyer wave, travels back in to the driver gas.

The interface, across which a limited degree of mixing occurs, separates driven and driver gases is referred to as the contact surface and follows, at a lower velocity, the lead wave.

A 'Chemical Shock Tube' involves separating driver and driven gases by a pair of diaphragms designed to fail after pre-determined delays with an end 'dump tank' of greatly increased cross-section. This allows an extreme rapid reduction (quench) in temperature of the heated gases.

Applications

In addition to measurements of rates of chemical kinetics shock tubes have been used to measure dissociation energies and molecular relaxation rates they have been used in aerodynamic tests. The fluid flow in the driven gas can be used much as a wind tunnel, allowing higher temperatures and pressures therein replicating conditions in the turbine sections of jet engines. However, test times are limited to a few milliseconds, either by the arrival of the contact surface or the reflected shock wave.

They have been further developed into shock tunnels, with an added nozzle and dump tank. The resultant high temperature hypersonic flow can be used to simulate atmospheric re-entry of space-craft or hypersonic craft, again with limited testing times.

Shock tubes have been developed in a wide range of sizes. The size and method of producing the shock wave determine the peak and duration of the pressure wave it produces. Thus, shock tubes can be used as a tool used to both create and direct blast waves at a sensor or an object in order to imitate actual explosions and the damage that they cause on a smaller scale. Results from shock tube experiments can be used to develop and validate numerical model of the response of a material or object to a blast wave. Shock tubes can be used to experimentally determine which materials and designs would be best suited to the job of attenuating blast waves. The results can then be incorporated into designs to protect structures and people that might be exposed to a blast wave. Shock tubes are also used in biomedical research to find out how biological tissues are affected by blast waves.

There are alternatives to the classical shock tube; for laboratory experiments at very high pressure, shock waves can also be created using high-intensity short-pulse lasers.

The shock tube figure (a) is a long tube of uniform cross section and with uniform internal dimensions. The diaphragm separates the high pressure driver section from the low pressure driven or test section. The material of the diaphragm and its thickness are dictated by the pressure ratio between the sections. On abrupt rupturing of the diaphragm, pressure waves emanating from the diaphragm station coalesce to form the shock front which propagates in to the low pressure section. As the shock front moves in to the low pressure section, a contact surface which is an imaginary line of separation between the driven and driver gases follows the shock front. The wave diagram of the shock tube after the diaphragm rupture is given in figure (b).

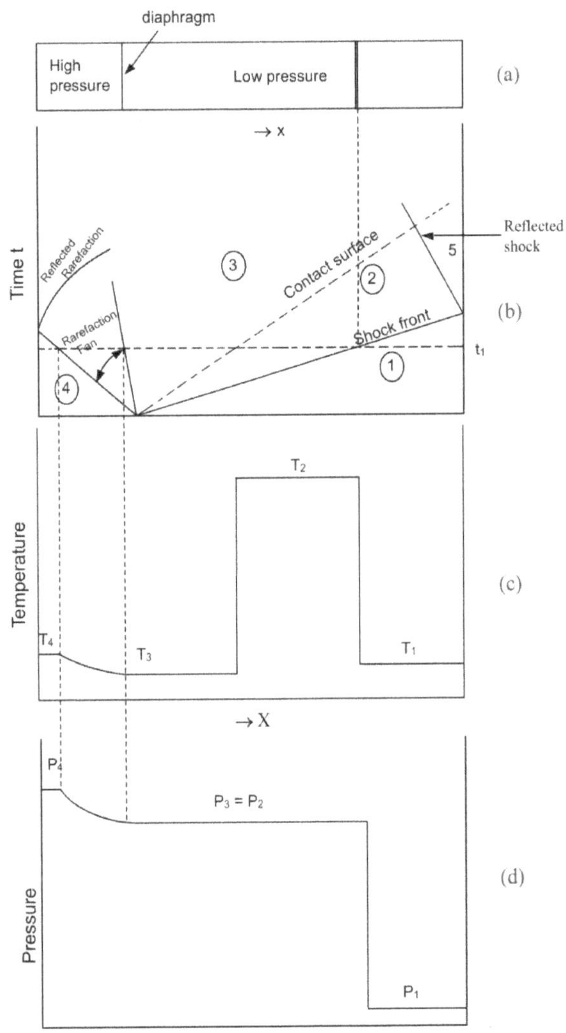

(a) The shock tube (b) The shock tube wave diagram
(c) Temperature history in the shock tube at time t_1
(d) Pressure history in the shock tube at time t_1

Simultaneously an expansion fan travels in to the driver section. The wave diagram in Fig. (b) shows these wave systems. The regions in the wave diagram are designated as 1, 2, 3 and 4. Region 1 represents the initial conditions in the driven section. The conditions of gas in this region are denoted by subscript 1.The Region 2 corresponds to the shocked gas conditions. The region between the contact surface and the expansion fan is referred to as 3. Region 4 represents the conditions of the initial high pressure driver gas.

The shock tube is a versatile experimental facility for the study of gaseous phenomena at elevated temperature and pressure. Along with the wave diagram, which is in the (x-t) plane, the diagrams showing the temperature and pressure history in the shock tube along its length are given in Fig. (c) and (d) at time. The temperature and pressure history corresponds to a particular time t_1 on the wave diagram. As shown in the figures, the experimental gas is brought almost instantly to a known and high temperature in region 2 and held at steady temperature and pressure for a few hundred micro sec. Usually there is some mixing and inter diffusion of gases at the contact surface so that the temperature fall is less sudden, than in the ideal case.

Shock Tube Equations

The equations applicable in the case of shock tube which are the moving normal shock equations may be derived considering the shock fixed coordinate system. In the figures, Ws represents the wave speed which is the speed at which the shock front propagates in the shock tube and v_1 and v_2 are the gas particle velocities ahead and behind the moving shock respectively. To convert the Laboratory fixed coordinates to shock fixed, wave speed Ws is applied in the opposite direction of the propagating shock.

$$u_1 = W_S - v_1 \text{ .... } 1$$
$$u_2 = W_S - v_2 \text{ .... } 2$$

u_1 and u_2 are the relative velocities in the shock fixed candidates

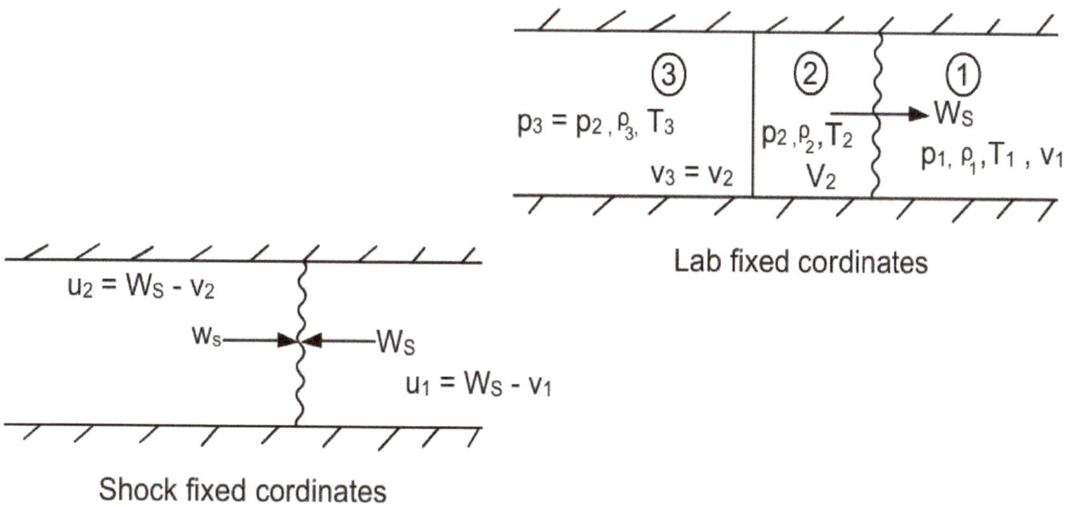

(a) Laboratory fixed coordinates (b) Shock fixed coordinates

Consider the conservation equations and the other thermodynamic equations as follows:

- *continuity* $\rho_1 u_1 = \rho_2 u_2$ 3
- *momentum* $p_1 + \rho_1 u_1^2 = p_2 + \rho_2 u_2^2$ 4
- *energy* $h_1 + \frac{1}{2}u_1^2 = h_2 + \frac{1}{2}u_2^2$ 5

$$h = C_p T$$

$$= \frac{\gamma R}{\gamma - 1} T = \frac{\gamma}{\gamma - 1}\frac{p}{\rho} \text{ .... } 6$$

Substituting in energy equation

$$\frac{\gamma}{\gamma - 1}\frac{p_1}{\rho_1} + \frac{1}{2}u_1^2 = \frac{\gamma}{\gamma - 1}\frac{p_p}{\rho_2} + \frac{1}{2}u_2^2$$

Using equations 3 to 6 u1 and u2 are eliminated from the equations $\frac{\rho_1 p_2 - \rho_2 p_1}{\rho_2 p_2 - \rho_1 p_1} = \frac{\gamma - 1}{\gamma + 1}$

Rearranging this equation we get

$$\frac{p_2}{p_1} = \frac{1 - \frac{\gamma - 1 \rho_1}{\gamma + 1 \rho_2}}{\frac{\rho_1}{\rho_2} - \frac{\gamma - 1}{\gamma + 1}} \quad \text{.....................7}$$

$$\frac{\rho_2}{\rho_1} = \frac{\frac{\gamma - 1}{\gamma + 1} + \frac{p_2}{p_1}}{\left(\frac{\gamma - 1}{\gamma + 1}\right)\frac{p_2}{p_1} + 1} \quad \text{.....................8}$$

The equations 7 and 8 are the Rankine - Hugoniot equations

From continuity and momentum equations

$$\frac{p_2}{p_1} = 1 + \frac{\rho_1 u_1^2}{p_1}\left(1 - \frac{\rho_1}{\rho_2}\right) \quad \text{.....................9}$$

$$Define\ M_1 = \frac{u_1}{c_1} = \frac{W_S}{c_1} \quad \text{.....................10}$$

Equation 10 is applicable when v_1 the particle velocity ahead of the moving shock is zero, in other words when shock is propagating in to still air

$$c_1 = \sqrt{\gamma R T_1} = \sqrt{\frac{\gamma p_1}{\rho_1}}$$

$$M_1 = \frac{u_1}{c_1}$$

Using equation 6 and 10

$$M_1 = u_1 \left(\frac{\rho_1}{\gamma p_1}\right)^{\frac{1}{2}}$$

$$M_1^2 = \frac{u_1^2 \rho_1}{\gamma p_1}$$

$$\gamma M_1^2 = u_1^2 \frac{\rho_1}{p_1} \quad \text{..................11}$$

Substitute equation 11 in 9 to write

$$\therefore \frac{p_2}{p_1} = 1 + \gamma M_1^2 \left(1 - \frac{\rho_1}{\rho_2}\right) \qquad \text{...........12}$$

Combining equations 8 and 12 gives

$$\frac{p_2}{p_1} = \frac{2\gamma M_1^2 - (\gamma - 1)}{(\gamma + 1)} \qquad \text{...........13}$$

$$\frac{\rho_2}{\rho_1} = \frac{(\gamma + 1M_1^2)}{(\gamma - 1)M_1^2 + 2} \qquad \text{...........14}$$

Combining the equations 13 and 14 with the perfect gas equation gives

$$\frac{T_2}{T_1} = \frac{\left(\gamma M_1^2 - \frac{\gamma - 1}{2}\right)\left(\frac{\gamma - 1}{2} M_1^2 + 1\right)}{\left(\frac{\gamma + 1}{2}\right)^2 M_1^2} \qquad \text{....................15}$$

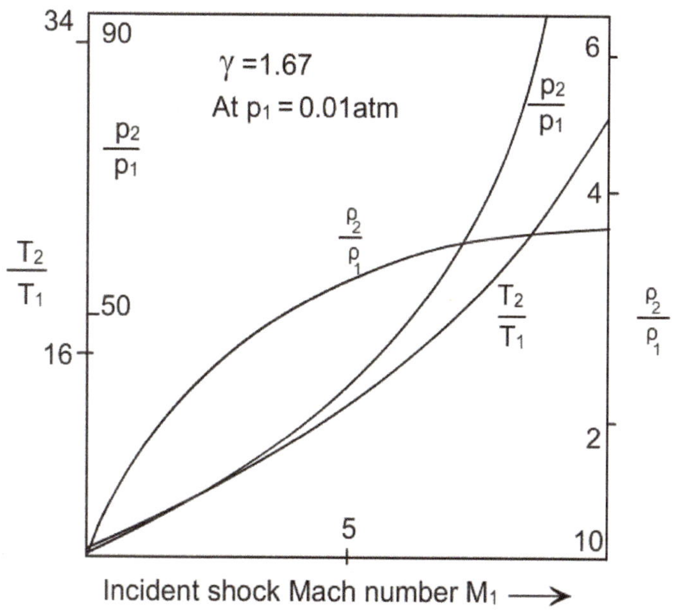

Theoretical ratios of shocked gas properties

Rankine–hugoniot Conditions

The Rankine–Hugoniot conditions, also referred to as Rankine–Hugoniot jump conditions or Rankine–Hugoniot relations, describe the relationship between the states on both sides of a shock wave in a one-dimensional flow in fluids or a one-dimensional deformation in solids. They are named in recognition of the work carried out by Scottish engineer and physicist William John Macquorn Rankine and French engineer Pierre Henri Hugoniot.

In a coordinate system that is moving with the shock, the Rankine–Hugoniot conditions can be expressed as:

$$\rho_1 u_s = \rho_2 (u_s - u_2) \qquad \text{Conservation of mass}$$
$$p_2 - p_1 = \rho_2 u_2 (u_s - u_2) = \rho_1 u_s u_2 \qquad \text{Conservation of momentum}$$
$$p_2 u_2 = \rho_1 u_s \left(\tfrac{1}{2} u_2^2 + e_2 - e_1\right) \qquad \text{Conservation of energy}$$

where u_s is the shock wave speed, ρ_1 and ρ_2 are the mass density of the fluid behind and inside the shock, u_2 is the particle velocity of the fluid inside the shock, p_1 and p_2 are the pressures in the two regions, and e_1 and e_2 are the *specific* (with the sense of *per unit mass*) internal energies in the two regions. These equations can be derived in a straightforward manner from equations (12), (13) and (14) below. Using the Rankine-Hugoniot equations for the conservation of mass and momentum to eliminate u_s and u_2, the equation for the conservation of energy can be expressed as the Hugoniot equation:

$$e_2 - e_1 = \tfrac{1}{2}(p_2 + p_1)(v_1 - v_2)$$

where v_1 and v_2 are the uncompressed and compressed specific volumes per unit mass, respectively.

Basics: Euler Equations in One Dimension

Consider gas in a one-dimensional container (e.g., a long thin tube). Assume that the fluid is inviscid (i.e., it shows no viscosity effects as for example friction with the tube walls). Furthermore, assume that there is no heat transfer by conduction or radiation and that gravitational acceleration can be neglected. Such a system can be described by the following system of conservation laws, known as the 1D Euler equations, that in conservation form is:

$$\frac{\partial \rho}{\partial t} = -\frac{\partial}{\partial x}(\rho u)$$

$$\frac{\partial}{\partial t}(\rho u) = -\frac{\partial}{\partial x}\left(\rho u^2 + p\right)$$

$$\frac{\partial}{\partial t}(E^t) = -\frac{\partial}{\partial x}\left[u\left(E^t + p\right)\right],$$

where

ρ = fluid mass density, [kg/m³]

u = fluid velocity, [m/s]

e = specific internal energy of the fluid, [J/kg]

p = fluid pressure, [Pa]

t = time, [s]

x = distance, [m], and

$E^t = \rho e + \rho \frac{1}{2} u^2$, is the total energy density of the fluid, [J/m³], while e is its specific internal energy [J/kg]

Assume further that the gas is calorically ideal and that therefore a polytropic equation-of-state of the simple form

$$p = (\gamma - 1)\rho e,$$

is valid, where γ is the constant ratio of specific heats c_p / c_v. This quantity also appears as the *polytropic exponent* of the polytropic process described by

$$\frac{p}{\rho^\gamma} = \text{constant.}$$

Note: For a calorically ideal gas γ is a constant and for a thermally ideal gas γ is a function of temperature. In the latter case, the dependence of pressure on mass density and internal energy might differ from that given by equation earlier.

The Jump Condition

Before proceeding further it is necessary to introduce the concept of a *jump condition* – a condition that holds at a discontinuity or abrupt change.

Consider a 1D situation where there is a jump in the scalar conserved physical quantity w, which is governed by integral conservation law

$$\frac{d}{dt}\int_{x_1}^{x_2} w dx = -f(w)\Big|_{x_1}^{x_2}$$

for any x_1, x_2, $x_1 < x_2$, and, therefore, by partial differential equation

$$\frac{\partial w}{\partial t} + \frac{\partial}{\partial x} f(w) = 0$$

for smooth solutions.

Let the solution exhibit a jump (or shock) at $x = x_s(t)$, where $x_1 < x_s(t)$ and $x_s(t) < x_2$, then

$$\frac{d}{dt}\left[\left(\int_{x_1}^{x_s(t)} w dx + \int_{x_s(t)}^{x_2} w dx\right)\right] = -\int_{x_1}^{x_2} \frac{\partial}{\partial x} f(w) dx$$

$$\therefore w_1 \frac{dx_s}{dt} - w_2 \frac{dx_s}{dt} + \int_{x_1}^{x_s(t)} w_t dx + \int_{x_s(t)}^{x_2} w_t dx = -f(w)\Big|_{x_1}^{x_2}$$

The subscripts *1* and *2* indicate conditions *just upstream* and *just downstream* of the jump respectively, i.e. $w_1 = \lim_{\in \to 0^+} w(x_s - \in)$ and $w_2 = \lim_{\epsilon \to 0^+} w(x_s + \in)$.

Note, to arrive at equation (8) we have used the fact that $dx_1 / dt = 0$ and $dx_2 / dt = 0$.

Now, let $x_1 \to x_s(t) - \in$ and $x_2 \to x_s(t) + \in$, when we have $\int_{x_1}^{x_s(t)-\epsilon} w_t dx \to 0$ and $\int_{x_s(t)+\grave{o}}^{x_2} w_t dx \to 0$, and in the limit

$$u_s(w_1 - w_2) = f(w_1) - f(w_2),$$

where we have defined $u_s = dx_s(t)/dt$ (the system *characteristic* or *shock speed*), which by simple division is given by

$$u_s = \frac{f(w_1) - f(w_2)}{w_1 - w_2}.$$

A shock situation arises in a system where its *characteristics* intersect, and under these conditions a requirement for a unique single-valued solution is that the solution should satisfy the *admissibility condition* or *entropy condition*. For physically real applications this means that the solution should satisfy the *Lax entropy condition*

$$f'(w_1) > u_s > f'(w_2),$$

where $f'(w_1)$ and $f'(w_2)$ represent *characteristic speeds* at upstream and downstream conditions respectively.

Euler Equations Shock Condition

In the case of the hyperbolic conservation law (6), we have seen that the shock speed can be obtained by simple division. However, for the 1D Euler equations, we have the vector state variable $[\rho, \rho u, E]^T$ and the jump conditions become

$$u_s(\rho_2 - \rho_1) = \rho_2 u_2 - \rho_1 u_1$$

$$u_s(\rho_2 u_2 - \rho_1 u_1) = (\rho_2 u_2^2 + p_2) - (\rho_1 u_1^2 + p_1)$$

$$u_s(E_2 - E_1) = \left[\rho_2 u_2\left(e_2 + \frac{1}{2}u_2^2 + p_2/\rho_2\right)\right] - \left[\rho_1 u_1\left(e_1 + \frac{1}{2}u_1^2 + p_1/\rho_1\right)\right].$$

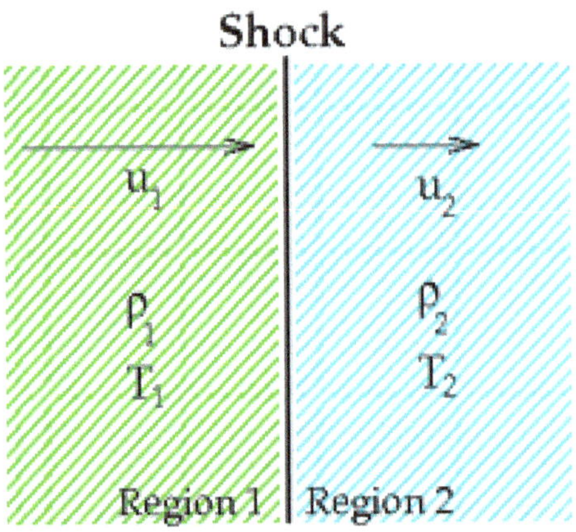

A schematic diagram of a shock wave situation with the density ρ, velocity u, and temperature T indicated for each region.

These equations are known as the *Rankine–Hugoniot conditions* for the Euler equations and are derived by enforcing the conservation laws in integral form over a control volume that includes the shock. For this situation u_s cannot be obtained by simple division. However, it can be shown by transforming the problem to a moving co-ordinate system (setting $u_s' := u_s - u_1$, $u_1' := 0$, $u_2' := u_2 - u_1$ to remove u_1) and some algebraic manipulation (involving the elimination of u_2' from the transformed equation, that the shock speed is given by

$$u_s = u_1 + c_1\sqrt{1 + \frac{\gamma+1}{2\gamma}\left(\frac{p_2}{p_1} - 1\right)},$$

where $c_1 = \sqrt{\gamma p_1 / \rho_1}$ is the speed of sound in the fluid at upstream conditions.

Stationary Shock

For a stationary shock $u_s = 0$, and for the 1D Euler equations we have

$$\rho_1 u_1 = \rho_2 u_2$$

$$\rho_1 u_1^2 + p_1 = \rho_2 u_2^2 + p_2$$

$$\rho_1 u_1\left(e_1 + \frac{1}{2}u_1^2 + p_1/\rho_1\right) = \rho_2 u_2\left(e_2 + \frac{1}{2}u_2^2 + p_2/\rho_2\right).$$

In view of equation earlier we can simplify equation above to

$$e_1 + \frac{1}{2}u_1^2 + p_1/\rho_1 = e_2 + \frac{1}{2}u_2^2 + p_2/\rho_2,$$

which is a statement of Bernoulli's principle, under conditions where gravitational effects can be neglected.

Substituting u_1 and u_2 from equations ealier into equation above yields the following relationship:

$$2(h_2 - h_1) = (p_2 - p_1)\cdot\left(\frac{1}{\rho_1} + \frac{1}{\rho_2}\right),$$

where $h = — + e$ represents specific enthalpy of the fluid. Eliminating internal energy e in equation by use of the equation-of-state, yields

$$\frac{\rho_2}{\rho_1} = \frac{\frac{p_2}{p_1}(\gamma+1) + (\gamma-1)}{(\gamma+1) + \frac{p_2}{p_1}(\gamma-1)} = \frac{u_1}{u_2}$$

$$\frac{p_2}{p_1} = \frac{\frac{\rho_2}{\rho_1}(\gamma+1)-(\gamma-1)}{(\gamma+1)-\frac{\rho_2}{\rho_1}(\gamma-1)}.$$

From physical considerations it is clear that both the upstream and downstream pressures must be positive, and this imposes an upper limit on the density ratio in equations above such that $\rho_2/\rho_1 < (\gamma+1)/(\gamma-1)$. As the strength of the shock increases, there is a corresponding increase in downstream gas temperature, but the density ratio ρ_2/ρ_1 approaches a finite limit: 4 for an ideal monatomic gas $(\gamma = 5/3)$ and 6 for an ideal diatomic gas $(\gamma = 1.4)$. Note: air is comprised predominately of diatomic molecules and therefore at standard conditions $\gamma_{air} \simeq 1.4$.

Shock Hugoniot and Rayleigh Line in Solids

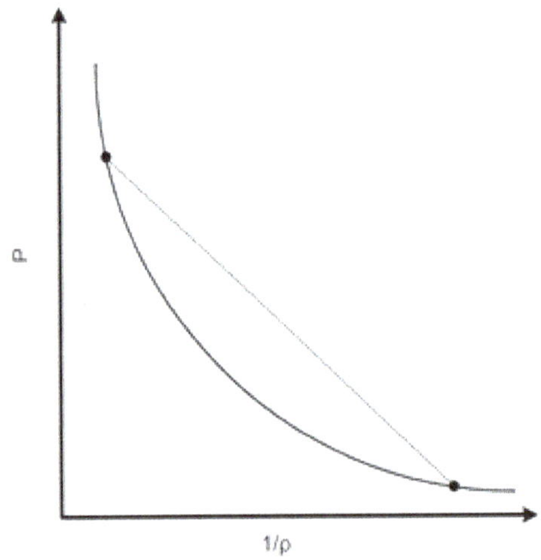

Shock Hugoniot and Rayleigh line in the *p-v* plane. The curve represents a plot of equation (24) with p_1, v_1, c_0, and *s* known. If $p_1 = 0$, the curve will intersect the specific volume axis at the point v_1.

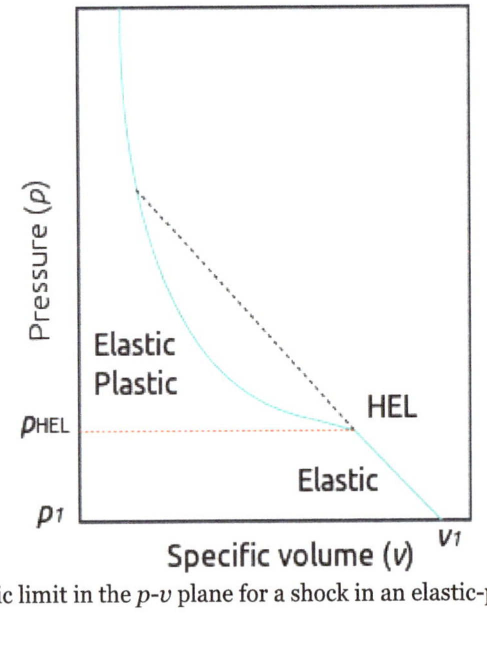

Hugoniot elastic limit in the *p-v* plane for a shock in an elastic-plastic material.

For shocks in solids, a closed form expression cannot be derived from first principles. Instead, experimental observations indicate that a linear relation can be used instead (called the shock Hugoniot in the u_s-u_p plane) that has the form

$$u_s = c_0 + s u_p = c_0 + s u_2$$

where c_0 is the bulk speed of sound in the material (in uniaxial compression), *s* is a parameter (the slope of the shock Hugoniot) obtained from fits to experimental data, and u_p=u_2 is the particle velocity inside the compressed region behind the shock front.

The above relation, when combined with the Hugoniot equations for the conservation of mass and momentum, can be used to determine the shock Hugoniot in the *p*-*v* plane, where *v* is the specific volume (per unit mass):

$$p_2 - p_1 = \frac{c_0^2 \rho_1 \rho_2 (\rho_2 - \rho_1)}{[\rho_2 - s(\rho_2 - \rho_1)]^2} = \frac{c_0^2 (v_1 - v_2)}{[v_1 - s(v_1 - v_2)]^2}.$$

Alternative equations of state, such as the Mie–Gruneisen equation of state may also be used instead of the above equation.

The shock Hugoniot describes the locus of all possible thermodynamic states a material can exist in behind a shock, projected onto a two dimensional state-state plane. It is therefore a set of equilibrium states and does not specifically represent the path through which a material undergoes transformation.

Weak shocks are isentropic and that the isentrope represents the path through which the material is loaded from the initial to final states by a compression wave with converging characteristics. In the case of weak shocks, the Hugoniot will therefore fall directly on the isentrope and can be used directly as the equivalent path. In the case of a strong shock we can no longer make that simplification directly. However, for engineering calculations, it is deemed that the isentrope is close enough to the Hugoniot that the same assumption can be made.

If the Hugoniot is approximately the loading path between states for an "equivalent" compression wave, then the jump conditions for the shock loading path can be determined by drawing a straight line between the initial and final states. This line is called the Rayleigh line and has the following equation:

$$p_2 - p_1 = u_s^2 \left(\rho_1 - \frac{\rho_1^2}{\rho_2} \right)$$

Hugoniot Elastic Limit

Most solid materials undergo plastic deformations when subjected to strong shocks. The point on the shock Hugoniot at which a material transitions from a purely elastic state to an elastic-plastic state is called the Hugoniot elastic limit (HEL) and the pressure at which this transition takes place is denoted p_{HEL}. Values of p_{HEL} can range from 0.2 GPa to 20 GPa. Above the HEL, the material loses much of its shear strength and starts behaving like a fluid.

Aerodynamic Heating

Aerodynamic heating is the heating of a solid body produced by its high-speed passage through air (or by the passage of air past a test object in a wind tunnel), whereby its kinetic energy is converted to heat by skin friction on the surface of the object at a rate that depends on the viscosity and speed of the air. In science and engineering, it is most frequently a concern regarding meteors, reentry vehicles, and the design of high-speed aircraft.

At high speeds through the air, the object's kinetic energy is converted to heat through compression and friction. At lower speed, the object will lose heat to the air through which it is passing, if the air is cooler. The combined temperature effect of heat from the air and from passage through it is called the stagnation temperature; the actual temperature is called the recovery temperature. These viscous dissipative effects to neighboring sub-layers make the boundary layer slow down via a non-isentropic process. Heat then conducts into the surface material from the higher temperature air. The result is an increase in the temperature of the material and a loss of energy from the flow. The forced convection ensures that other material replenishes the gases that have cooled to continue the process.

The stagnation and the recovery temperature of a flow increases with the speed of the flow and are greater at high speeds. The total thermal loading of the object is a function of both the recovery temperature and the mass flow rate of the flow. Aerodynamic heating is greatest at high speed and in the lower atmosphere where the density is greater. In addition to the convective process described above, there is also Thermal radiation from the flow to the body and vice versa with the net direction set by the relative temperature of each.

Aerodynamic heating increases with the speed of the vehicle. Its effects are minimal at subsonic speeds but at supersonic speeds beyond about M2.2 it dictates the design/materials of the vehicle structure and internal systems. The heating effects are greatest at leading edges but the whole vehicle heats up to a stabilized temperature if it remains at speed. Aerodynamic heating is dealt with by the use of high temperature alloys for metals, the addition of insulation of the exterior of the vehicle, or the use of ablative material.

Aircraft

Aerodynamic heating is a concern for supersonic and hypersonic aircraft. The SR-71 used titanium skin panels painted black to reduce the temperature and corrugated to accommodate expansion. Some designs for hypersonic missiles have used liquid cooling of the leading edges (usually the fuel en route to the engine). The Sprint missile's heat shield needed several design iterations for Mach 10 temperatures.

Reentry Vehicles

Heating caused by the very high reentry speeds (greater than Mach 20) is sufficient to destroy the vehicle unless special techniques are used. The early space capsules such as used on Mercury, Gemini, and Apollo were given blunt shapes to produce a stand-off bow shock. As a result most of the heat is dissipated to surrounding air without transferring through the vehicle structure. Additionally, these vehicles had ablative material that sublimates into a gas at high temperature. The act of sublimation absorbs the thermal energy from the aerodynamic heating and erodes the material away as opposed to heating the capsule. The surface of the heat shield for the Mercury spacecraft had a coating of aluminum with glassfiber in many layers. As the temperature rose to 1,100 °C (1,400 K) the layers would evaporate and take the heat with it. The spacecraft would become hot but not harmfully so. The Space Shuttle used insulating tiles on its lower surface to absorb and radiate heat while preventing conduction to the aluminum airframe. The damage to the heat shield during liftoff of Space Shuttle *Columbia* contributed to its destruction upon reentry.

Comparison Between Shock Heating and Isentropic Heating

For a given pressure change (or density change) in a gas, the temperature produced by a strong shock wave is greatly in excess of those in the isentropic case. Shock wave is an irreversible adiabatic compression

$$h_1 + \frac{1}{2}v_1^2 = h_2 + \frac{1}{2}v_2^2$$

Conservation of energy derived for flow where a resistance to motion occurs. In the case of shock wave, the abrupt pressure rise across the front presents a resistance to flow and as such implies an irreversible conversion of kinetic energy (K.E) into heat.

$\left(\frac{1}{2}v_1^2 - \frac{1}{2}v_2^2\right)$ *is a measure of this change in K.E for unit mass of gas.*

Across the shock front there is an increase of entropy, for an adiabatic process.

$$\Delta S = \int \frac{dQ}{T} = c_v \int_{T_1}^{T_2} \frac{dT}{T} + \int_{T_1}^{T_2} p \frac{dv}{T}$$

From this, for a perfect gas

$$\Delta S = C_p \; In \left\{ \frac{T_2 / T_1}{(p_2 / p_1)^{\frac{\gamma-1}{\gamma}}} \right\}$$

$$\left(\frac{T_2}{T_1}\right) = \left(\frac{p_2}{p_1}\right)^{\frac{\gamma-1}{\gamma}} \exp\left(\frac{\Delta s}{c_P}\right) \qquad \text{.........................16}$$

From this it can be seen that (T2/T1) across a shock front ($\Delta S > 0$) exceeds that for an isentropic compression ($\Delta S = 0$) by a factor $\left[\exp\left(\frac{\Delta S}{C_P}\right)\right]$ between the same limits. Compression by a piston across which differential pressure is kept small.

Table: Comparison between isentropic and shock heating

$P_2 P_1$	Shock Wave (T_2 K)	Isentropic Compression (T_2 K)	M_1
2	420	357	1.32
10	895	566	2.95
25	1753	735	4.45
50	3177	897	6.55

Particle Velocity Behind the Moving Shock Wave

Refer to Figure (b) of shock fixed coordinates

$$u_1 = W_S - v_1$$
$$u_2 = W_S - v_2$$
$$u_1 - u_2 = -(v_1 - v_2)$$
$$= v_2 - v_1$$

$$\frac{\rho_2}{\rho_1} = \frac{(\gamma+1)M_1^2}{(\gamma-1)M_1^2+2} = \frac{u_1}{u_2} = \frac{W_S - v_1}{W_S - v_2}$$

$$\frac{\rho_2}{\rho_1} = \frac{\gamma+1}{2}\frac{M_1^2}{1+\frac{\gamma-1}{2}M_1^2}$$

$$u_1 = M_1 c_1$$

$$u_2 = u_1\left[\frac{(\gamma-1)M_1^2+2}{(\gamma+1)M_1^2}\right]$$

$$u_2 = M_1 c_1\left[\frac{(\gamma-1)M_1^2+2}{(\gamma+1)M_1^2}\right]$$

$$u_2 - u_1 = c_1\left[\frac{(\gamma-1)M_1^2+2}{(\gamma+1)M_1} - M_1\right]$$

$$= c_1\left[\frac{\gamma M_1^2 - M_1^2 + 2 - \gamma M_1^2 - M_1^2}{(\gamma+1)M_1}\right] = \frac{c_1 \times 2\left(1-M_1^2\right)}{(\gamma+1)M_1}$$

$$u_1 - u_2 = \frac{2c_1}{\gamma+1}\left(\frac{M_1^2-1}{M_1}\right)$$

$$\frac{2c_1}{\gamma+1}\left(M_1 - \frac{1}{M_1}\right)$$

$$v_2 = \frac{2c}{\gamma+1}\left(M_1 - \frac{1}{M_1}\right) + v_1 \quad \ldots\ldots\ldots\ldots\ldots\ldots 17$$

Illustration of Calculation of Particle Velocity (v_2) and Mach Number Behind the Moving Shock

Calculations are done assuming Air to be the test gas at 300K

- For $M_1 = 8$

$T_2 / T_1 = 13.387$ (from Gas Tables)

$T_2 = 4016.1$ $\quad$ $c_2 = 1270$m/s

$$v_2 = \frac{2c_1}{\gamma+1}\left(M_1 - \frac{1}{M_1}\right) = 2280 m/s$$

- $M_2 = 1.79$

- *For* $M_1 = 10$

$\frac{T_2}{T_1} = 20.388$ (From Gas Tables)

$T_2 = 6116.4$ $\quad$ $c_2 = 1567$m/s

$$v_2 = \frac{2\times 347.5}{2.4}(10-0.1) = 2866\text{m/s} = 1.829$$

- For $M_1 = 25$

$$\frac{T_2}{T_1} = \frac{\gamma\left(\gamma M_1^2 - \frac{\gamma-1}{2}\right)\left(\frac{\gamma-1}{2}M_1^2+1\right)}{\left(\frac{\gamma+1}{2}\right)^2 M_1^2}$$

$$= \frac{(1.4\times 625 - 0.2)(0.2\times 625+1)}{1.44\times 625}$$

$$= \frac{(874.8)(126)}{900} = 122.47$$

- $T_2 = 36741$

$$\sqrt{T_2} = 191.68 \times 20.04$$

$$C_2 = 3842.21 m/s$$

$$v_2 = \frac{2 \times 347.5}{2.4}\left(25 - \frac{1}{25}\right) \times 24.96$$

$$\therefore \frac{7228}{3842} = 1.88$$

What is shown above is that increase of shock velocity (Ws) and consequently the shock Mach number will not correspondingly raise the Mach number after the shock (M_2) based on v_2. Understandably, the reason is the increase in the post shock temperature (T_2) and the value of speed of sound (c_2) after the shock to high values.

Dependence of Shock Strength on Diaphragm Pressure Ratio

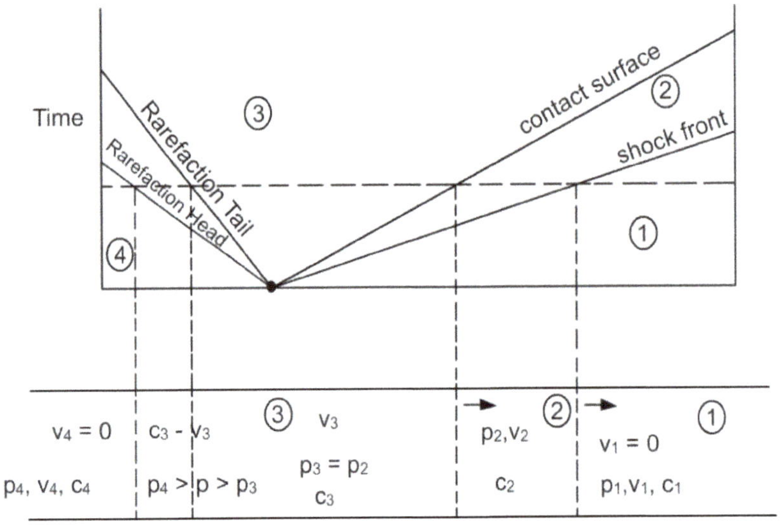

Wave diagram and the gas properties along the shock tube at time t_1

Driver gas at a high pressure p_4 in region 4 is being expanded through the expansion fan to a lower pressure p_3 behind the contact surface. Limits of expansion fan are formed by the head and tail of rarefaction wave.

Expansion is an isentropic process. Applying Newton's 2nd law, $F \quad m\frac{dv}{dt}$ to a one dimensional fluid element of thickness dx. Under the action of a sound wave travelling at a velocity c, we get

$$-dp = \rho dx \frac{dv}{\frac{1}{c}dx} = \rho c dv \qquad \text{.......................18}$$

$$\rho dx \rightarrow m$$

$$\frac{1}{c}dx \rightarrow dt$$

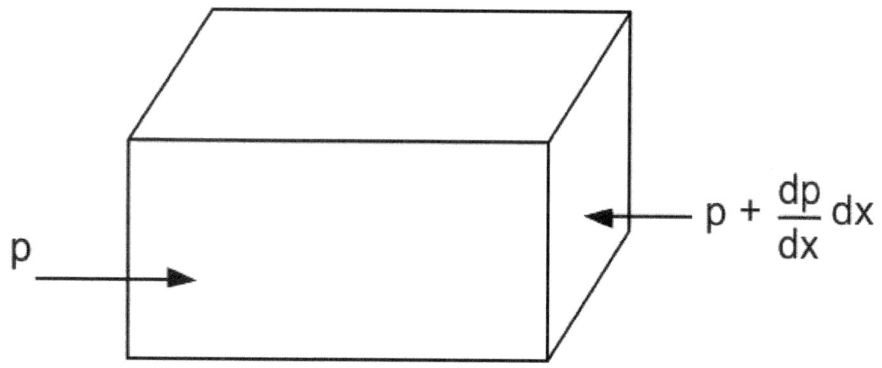

Elemental fluid volume subjected to forces

If an ideal gas with constant specific heats undergoes a change of state the entropy change ΔS of the gas is

$$\Delta S = c_p \ln\frac{T_f}{T_i} - R \ln\frac{p_f}{p_i}$$

In differential form $dS = c_P \frac{dT}{T} - R\frac{dp}{p}$

For an isentropic process ds = 0

Hence, $\frac{dp}{p} = \frac{c_P}{R}\frac{dT}{T}$ 19

$$c^2 = \gamma RT \quad ie \quad T = \frac{c^2}{\gamma R}$$

$$dT = \frac{2cdc}{\gamma R}$$

$$\frac{dT}{T} = \frac{2cdc}{\gamma R \frac{c^2}{\gamma R}} = \frac{2dc}{c} \quad ....20$$

Combining (19) and (20)

$$\frac{dp}{p} = \frac{2c_p dc}{Rc} \quad ....21$$

From Equation (18) $-dp = \rho \, c \, dV$

From (4) $-\frac{\rho c dv}{p} = \frac{2c_p dc}{Rc}$

$$2c_p dc = \frac{\rho c^2 R dv}{p} = \frac{\rho \gamma p}{\rho}\frac{R dv}{p}$$

$$2c_p dc = 2\frac{\gamma R}{\gamma - 1}dc = -\gamma R dv$$

$$C_P = \frac{\gamma R}{\gamma - 1} \; ie \frac{2}{\gamma - 1}dc + dv = 0$$

Thus, for an isentropic process

$$\left(\frac{2}{\gamma - 1}\right)c + v = \text{constant}$$

For the shock tube problem

$$\left(\frac{2}{\gamma_4 - 1}\right)c_4 + v_4 = \left(\frac{2}{\gamma_3 - 1}\right)c_3 + v_3$$

On each side of the contact surface, the particle velocities and pressures must be continuous, since there is no gas flow across the interface.

$$v_3 = v_2 \; and \; p_3 = p_2$$

$$As \; v_4 = 0 \; and \; \gamma_4 = \gamma_3$$

$$\frac{2}{\gamma_4 - 1}c_4 = \frac{2}{\gamma_4 - 1}c_3 + v_2$$

$$c_3 = c_4 - \frac{\gamma_{4-1}}{2}v_2$$

From adiabatic gas law

$$\frac{p_4}{p_3} = \left(\frac{c_4}{c_3}\right)^{\frac{2\gamma_4}{\gamma_{4-1}}} = \frac{p_4}{p_2} \qquad P_3 = P_2$$

$$\therefore \frac{p_4}{p_2} = \left(\frac{c_4}{c_4 - \frac{\gamma_{4-1}}{2}v_2}\right)^{\frac{2\gamma_4}{\gamma_{4-1}}} \qquad \text{..............................22}$$

$$v_2 = \frac{2c_1}{\gamma_{1+1}}\left(M_1 - \frac{1}{M_1}\right) \qquad \text{..........................23}$$

Combining these equations with equation for $\frac{p_2}{p_1} = \frac{2\gamma M_1^2 - \gamma - 1}{\gamma + 1}$ leads to

$$\frac{p_4}{p_1} = \frac{2\gamma_1 M_1^2 - (\gamma_1 - 1)}{(\gamma_1 + 1)} \left\{ 1 - \frac{\gamma_{4-1}}{\gamma_{1+1}} \frac{c_1}{c_4} \left(M_1 - \frac{1}{M_1} \right) \right\}^{-\left(\frac{2\gamma_4}{\gamma_{4-1}}\right)} \quad \text{..............................24}$$

Strongest shocks are obtained when

$\frac{p_4}{p_1} \to \infty$ and then

$$\left\{ 1 - \frac{\gamma_{4-1}}{\gamma_{1+1}} \frac{c_1}{c_4} \left(M_1 - \frac{1}{M_1} \right) \right\} \to 0 \qquad M_1 \to \frac{(\gamma_1 + 1)}{(\gamma_4 - 1)} \frac{c_4}{c_1}$$

Strongest shocks are obtained for driver with

1. High speed of sound
2. Low specific heat ratio

Reflected Shocks

When the primary shock wave reaches the end of the shock tube it gets reflected back in to the medium which is already heated by the incident shock. Higher temperature and pressure are obtained behind the reflected shock. Additionally the gas behind the reflected wave is at rest relative to the shock tube.

Summarizing:

- Much higher temperatures are obtained that too at high pressures. Dissociation is prevented because of increased pressures.
- Study on a fixed group of molecules can be conducted. Behind the incident shock, the gas particles are moving with a velocity v_2 and it is impossible to follow a fixed volume of gas.

Reflected Shock Parameters

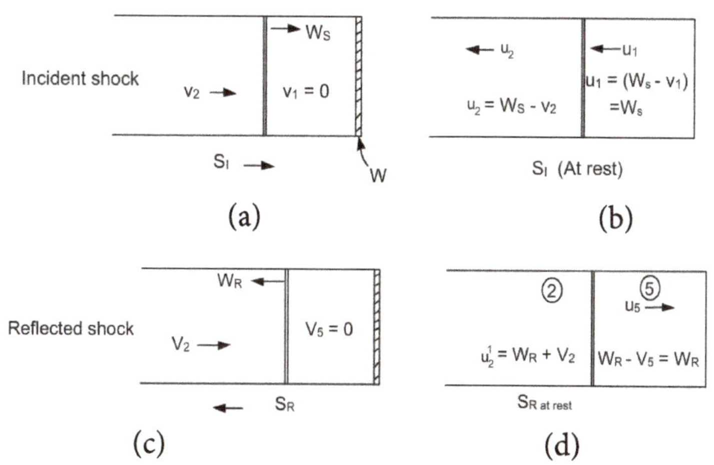

Incident and reflected shocks: (a&c) Laboratory fixed coordinates, (b&d) Shock fixed coordinates

The subscript 5 refers to the reflected gas properties. The gas flows into the shock front with a relative velocity u'_2 where

$$u'_2 = W_R + v_2$$

In the reflected shock region $v_5 = 0$. This means that the gas gives all its kinetic energy (K.E) on passing through the front thus increasing the properties of the state of the gas. This can be seen as an ever extending column which advances out from the end wall of the tube and contains a gas at high temperature; density and pressure. The particle velocities relative to the reflected shock at different sections can be written.

$$u_5 = W_R - v_5 \equiv W_R$$
$$u_1^2 = W_R + v_2$$

If Mach no. of the reflected shock is defined as

$$M_R = \frac{u'_2}{c_2}$$

Equations relating to reflected shock may be written analogy with those with incident shock. For incident shock

$$u_1 - u_2 = v_2 - v_1 = \frac{2c_1}{\gamma+1}\left(M_1 - \frac{1}{M_1}\right)$$
$$u_1 = W_S$$
$$u_2 = W_S - v_2$$
$$u'_2 - u_5 = v_2 + v_5^0$$
$$= \frac{2c_2}{\gamma+1}\left(M_R - \frac{1}{M_R}\right)$$
$$v_2 = \frac{2c_1}{\gamma+1}\left(M_1 - \frac{1}{M_1}\right) = \frac{2c_2}{\gamma+1}\left(M_R - \frac{1}{M_R}\right) \quad \text{......................25}$$

From perfect gas equation and equation of the speed of sound

$$\left(\frac{c_2}{c_1}\right)^2 = \frac{p_2}{p_1} \times \frac{\rho_1}{\rho_2} \quad \text{..................................26}$$

From Rankine -Hugoniot equations

$$\frac{\rho_2}{\rho_1}=\frac{\dfrac{\gamma-1}{\gamma+2}+\dfrac{p_2}{p_1}}{\left(\dfrac{\gamma-1}{\gamma+1}\right)\dfrac{p_2}{p_1}+1}$$

$$\frac{p_T}{\rho_1}=RT_1$$

$$\frac{p_2}{\rho_2}=RT_2$$

$$\left(\frac{c_2}{c_1}\right)^2=\frac{p_2}{p_1}\left(1+\frac{\gamma-1}{\gamma+1}\frac{p_2}{p_1}\right)\left(\frac{\gamma-1}{\gamma+1}+\frac{p_2}{p_1}\right)^{-1} \quad \text{..................................27}$$

Substitute in the incident shock equations

$$\therefore \frac{p_2}{p_1}=\frac{2\gamma M_1^2-(\gamma-1)}{\gamma+1}$$

By analogy, we may write for reflected shock

$$\frac{p_5}{p_2}=\frac{2\gamma M_R^2-(\gamma-1)}{\gamma+1} \quad \text{........................28}$$

Solving equations (25) to (28)

$$\left(\frac{p_5}{p_2}\right)=\frac{\dfrac{\gamma+1}{\gamma-1}+2-\dfrac{p_1}{p_2}}{1+\dfrac{\gamma+1}{\gamma-1}\dfrac{p_1}{p_2}} \quad \text{........................29}$$

$\left(\dfrac{\rho_5}{\rho_2}\right)$ can be written by analogy

$$\left(\frac{p_5}{p_2}\right)=\frac{\dfrac{\gamma-1}{\gamma+1}+\dfrac{p_5}{p_2}}{\left(\dfrac{\gamma-1}{\gamma+1}\right)\dfrac{p_5}{p_2}+1} \quad \text{........................30}$$

Similarly,

$$\frac{T_5}{T_2}=\left(\frac{p_5}{p_2}\right)\left\{\frac{\dfrac{\gamma+1}{\gamma-1}+\dfrac{p_5}{p_2}}{\left(\dfrac{\gamma+1}{\gamma-1}\right)\dfrac{p_5}{p_2}}\right\} \quad \text{........................31}$$

These ratios can also be written as:

$$\frac{}{p_1}\left[\frac{2\gamma \qquad -(\gamma-1)}{\gamma+1}\right]\left[\frac{(3\gamma-1) \qquad -2(\gamma-1)}{(\gamma-1)M_1 \quad +2}\right]$$

$$\frac{T_5}{T_1}\frac{\left[2(\gamma-1)M_1^2+(3-\gamma)\right]\left[(3\gamma-1)M_1^2-2(\gamma-1)\right]}{(\gamma+1)M_1^2}$$

The ratio of speeds can be written as

$$\frac{W_R}{W_S}=\frac{2+\dfrac{2}{\gamma-1}\dfrac{p_1}{p_2}}{\dfrac{\gamma+1}{\gamma-1}-\dfrac{p_1}{p_2}}$$

Table: Reflected shock parameters in terms of incident parameters

γ	M_1	$\frac{p_2}{p_1}$	$\frac{p_5}{p_2}$	$\frac{T_2}{T_1}$	$\frac{T_5}{T_2}$	$\frac{W_R}{W_S}$
1.4	2.95	10	4.95	2.62	1.76	0.423
	6.56	50	7.12	9.31	2.28	0.351
1.66	2.87	10	4.22	3.42	1.94	0.589
	6.34	50	5.54	13.4	2.34	0.517

Viscous Effects and the Shock Tube Boundary Layer

Due to the viscous nature of the flow in the shock tube, boundary layer is formed. Its thickness will be zero at the shock front and increases back through the shock heated region and the contact surface in to the expanding driver gas and becomes zero again at the head of the rarefaction fan. Schematic of the boundary layer formed is shown in figure. Important effects of formation of the boundary layer are the following.

- Kinetic energy is dissipated as heat in the retarding layer of the boundary layer and this is conveyed to walls as by heat transfer.
- Deceleration of the shock front.
- Acceleration of contact surface.

One of the important effects of the formation of boundary layer is its reducing effects of the useful observation time in shock tube experiments.

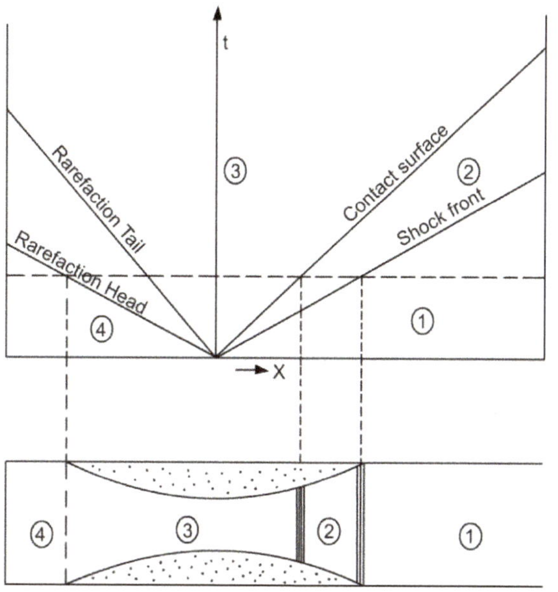

Schematic of the growth of boundary layer in a shock tube

Observation Time in Shock Tube

Observation time is decided by

(a) length of observation station from diaphragm

(b) length of driver section [due to the influence reflected rarefaction fan]

(c) length of end plate from diaphragm [influence of the reflected shock]

(d) growth of the boundary layer

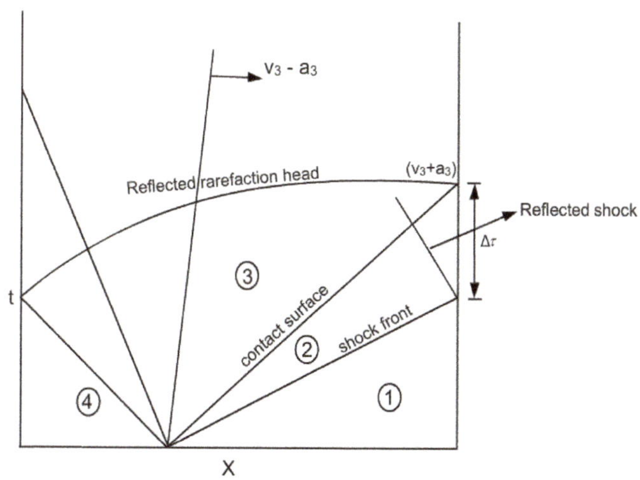

(x,t) diagram of the shock tube indicating the different waves

Interaction of Reflected Shock and Contact Surface

The reflected shock, after going through the shocked gas in the region 2 encounters the contact surface and the reaction will be dependent on the speed.

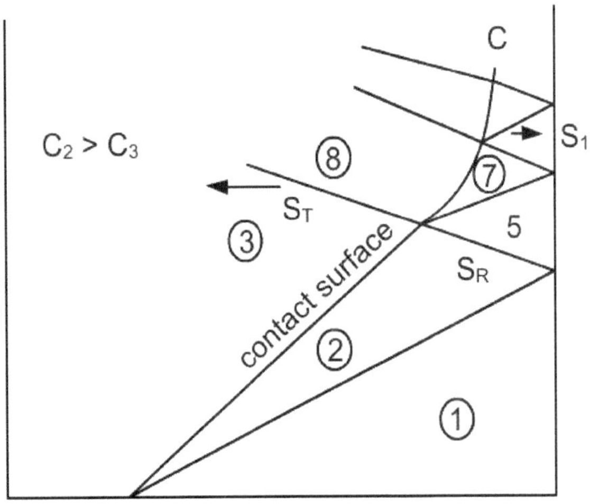

(a)

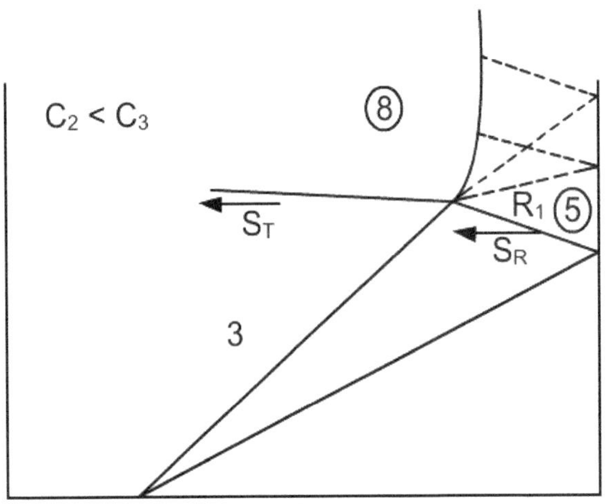

(b)

Time available for measurements in the reflected shock region can be improved.

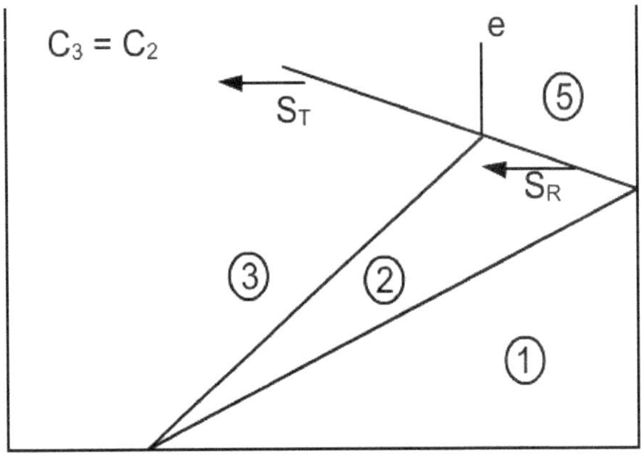

(c)

Interaction of reflected shock with the contact surface (a) $c_2 > c_3$ (b) $c_2 < c_3$ (c) $c_2 = c_3$

Speed of sound in regions 2 and 3. The schematic of the wave systems generated based on the values of c_2 and c_3 is shown in Figures (a) and (b). As shown in Fig. (a) when $c_2 > c_3$, the reflected shock after passing through the contact surface enters a region of higher Mach number. The resultant properties in the region 3 would be greater than those in regions 2.This is physically not feasible and in order to make the pressures on either side of the contact surface an additional shock system is generated in region 5. The opposite effect is there corresponding to the situation in Fig. (b) where $c_2 < c_3$.The resultant wave system in the reflected shock region is an expansion wave. In both the cases, flow in the reflected shock region is disturbed by the additional waves generated. In order to have larger observation times behind the reflected shock c_2 is designed to be same as c_3.This is called 'tailoring' of the shock tube contact surface. The wave system when $c_2 = c_3$ is shown in Fig. (c).Tailoring enhances the observation time behind the reflected shock.

Shock Tube Diaphragm and Bursting Techniques

Diaphragms vary from 0.025mm cellophane sheet to 6mm thick steel plate. For the same material the diaphragm rupturing pressure increases directly as the pressure and inversely as the exposed area.

- pressure driving
- Mechanical piston drive
- Combustion driving
- Heating of driver gas

Measurement of Shock Speed

Conventionally, the time distance method is used for the measurement of shock speed. Sensors of pressure or temperature are mounted flush with the interior of the low pressure section of the shock tube at known distance between them. Time taken by the propagating shock to transverse the known distance is used to calculate the shock speed. Though this is a simple method to implement, the speed obtained is an average value between the two stations. Other methods such as that making use of Doppler principle could give higher spatial resolution.

Mach Reflection

Mach reflection is a supersonic fluid dynamics effect, named for Ernst Mach, and is a shock wave reflection pattern involving three shocks.

Introduction

Mach reflection can exist in steady, pseudo-steady and unsteady flows. When a shock wave, which is moving with a constant velocity, propagates over a solid wedge, the flow generated by the shock impinges on the wedge thus generating a second reflected shock, which ensures that the velocity of the flow is parallel to the wedge surface. Viewed in the frame of the reflection point, this flow is locally steady, and the flow is referred to as pseudosteady. When the angle between the wedge and the primary shock is sufficiently large, a single reflected shock is not able to turn the flow to a direction parallel to the wall and a transition to Mach reflection occurs.

In a steady flow situation, if a wedge is placed into a steady supersonic flow in such a way that its oblique attached shock impinges on a flat wall parallel to the free stream, the shock turns the flow toward the wall and a reflected shock is required to turn the flow back to a direction parallel to the wall. When the shock angle exceeds a certain value, the deflection achievable by a single reflected shock is insufficient to turn the flow back to a direction parallel to the wall and transition to Mach reflection is observed.

Mach reflection consists of three shocks, namely the incident shock, the reflected shock and a Mach stem, as well as a slip plane. The point where the three shocks meet is known as the 'triple point' in two dimensions, or a shock-shock in three dimensions.

Types of Mach Reflection

The only type of Mach reflection possible in steady flow is direct-Mach reflection, in which the Mach stem is convex away from the oncoming flow, and the slip plane slopes towards the reflecting surface.

By new results there is a new configuration of shock waves - configuration with a negative angle of reflection in steady flow.

In Pseudo-steady flows, the triple point moves away from the reflecting surface and the reflection is a direct-Mach reflection. In unsteady flows, it is also possible that the triple point remains stationary relative to the reflecting surface (stationary-Mach reflection), or moves toward the reflecting surface (inverse-Mach reflection). In inverse Mach reflection, the Mach stem is convex toward the oncoming flow, and the slip plane curves away from the reflecting surface. Each one of these configurations can assume one of the following three possibilities: single-Mach reflection, transitional-Mach reflection and double-Mach reflection.

Compressible Flow

Compressible flow (gas dynamics) is the branch of fluid mechanics that deals with flows having significant changes in fluid density. Gases, mostly, display such behaviour. While all flows are compressible, flows are usually treated as being incompressible when the Mach number (the ratio of the speed of the flow to the speed of sound) is less than 0.3 (since the density change due to velocity is about 5% in that case). The study of compressible flow is relevant to high-speed aircraft, jet engines, rocket motors, high-speed entry into a planetary atmosphere, gas pipelines, commercial applications such as abrasive blasting, and many other fields.

History

The study of gas dynamics is often associated with the flight of modern high-speed aircraft and atmospheric reentry of space-exploration vehicles; however, its origins lie with simpler machines. At the beginning of the 19th century, investigation into the behaviour of fired bullets led to improvement in the accuracy and capabilities of guns and artillery. As the century progressed, inventors such as Gustaf de Laval advanced the field, while researchers such as Ernst Mach sought to understand the physical phenomenon involved through experimentation.

At the beginning of the 20th century, the focus of gas dynamics research shifted to what would eventually become the aerospace industry. Ludwig Prandtl and his students proposed important concepts ranging from the boundary layer to supersonic shock waves, supersonic wind tunnels, and supersonic nozzle design. Theodore von Kármán, a student of Prandtl, continued to improve the understanding of supersonic flow. Other notable figures (Meyer, Luigi Crocco, and Shapiro) also contributed significantly to the principles considered fundamental to the study of modern gas dynamics. Many others also contributed to this field.

Accompanying the improved conceptual understanding of gas dynamics in the early 20th century was a public misconception that there existed a barrier to the attainable speed of aircraft, commonly referred to as the "sound barrier." In truth, the barrier to supersonic flight was merely a technological one, although it was a stubborn barrier to overcome. Amongst other factors, conventional aerofoils saw a dramatic increase in drag coefficient when the flow approached the speed of sound. Overcoming the larger drag proved difficult with contemporary designs, thus the perception of a sound barrier. However, aircraft design progressed sufficiently to produce the Bell X-1. Piloted by Chuck Yeager, the X-1 officially achieved supersonic speed in October 1947.

Historically, two parallel paths of research have been followed in order to further gas dynamics knowledge. Experimental gas dynamics undertakes wind tunnel model experiments and experiments in shock tubes and ballistic ranges with the use of optical techniques to document the findings. Theoretical gas dynamics considers the equations of motion applied to a variable-density gas, and their solutions. Much of basic gas dynamics is analytical, but in the modern era Computational fluid dynamics applies computing power to solve the otherwise-intractable nonlinear partial differential equations of compressible flow for specific geometries and flow characteristics.

Introductory Concepts

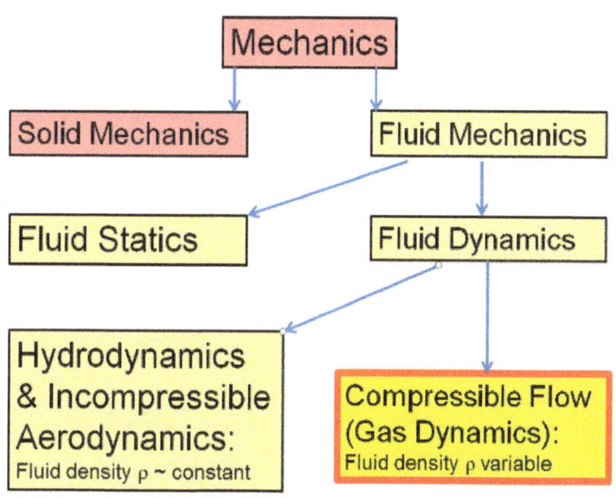

Breakdown of fluid mechanics chart

There are several important assumptions involved in the underlying theory of compressible flow. All fluids are composed of molecules, but tracking a huge number of individual molecules in a flow is unnecessary. Instead, the continuum assumption allows us to consider a flowing gas as a continuous substance except at low densities. This assumption provides a huge simplification which is accurate for most gas-dynamic problems. Only in the low-density realm of rarefied gas dynamics does the motion of individual molecules become important.

A related assumption is the no-slip condition where the flow velocity at a solid surface is presumed equal to the velocity of the surface itself, which is a direct consequence of assuming continuum flow. The no-slip condition implies that the flow is viscous, and as a result a boundary layer forms on bodies traveling through the air at high speeds, much as it does in low-speed flow.

Most problems in incompressible flow involve only two unknowns: pressure and velocity, which are typically found by solving the two equations that describe conservation of mass and of linear momentum, with the fluid density presumed constant. In compressible flow, however, the gas density and temperature also become variables. This requires two more equations in order to solve compressible-flow problems: an equation of state for the gas and a conservation of energy equation. For the majority of gas-dynamic problems, the simple Ideal gas law is the appropriate state equation.

Fluid dynamics problems have two overall types of references frames, called Lagrangian and Eulerian. The Lagrangian approach follows a fluid mass of fixed identity as it moves through a flowfield. The Eulerian reference frame, in contrast, does not move with the fluid. Rather it is a fixed frame or control volume that fluid flows through. The Eulerian frame is most useful in a majority of compressible flow problems, but requires that the equations of motion be written in a compatible format.

Finally, although space is known to have 3 dimensions, an important simplification can be had in describing gas dynamics mathematically if only one spatial dimension is of primary importance, hence 1-dimensional flow is assumed. This works well in duct, nozzle, and diffuser flows where the flow properties change mainly in the flow direction rather than perpendicular to the flow. However, an important class of compressible flows, including the external flow over bodies traveling at high speed, requires at least a 2-dimensional treatment. When all 3 spatial dimensions and perhaps the time dimension as well are important, we often resort to computerized solutions of the governing equations.

Mach Number, Wave Motion, and Sonic Speed

The Mach number (M) is defined as the ratio of the speed of an object (or of a flow) to the speed of sound. For instance, in air at room temperature, the speed of sound is about 340 m/s (1,100 ft/s). M can range from 0 to ∞, but this broad range falls naturally into several flow regimes. These regimes are subsonic, transonic, supersonic, hypersonic, and hypervelocity flow. The figure below illustrates the Mach number "spectrum" of these flow regimes.

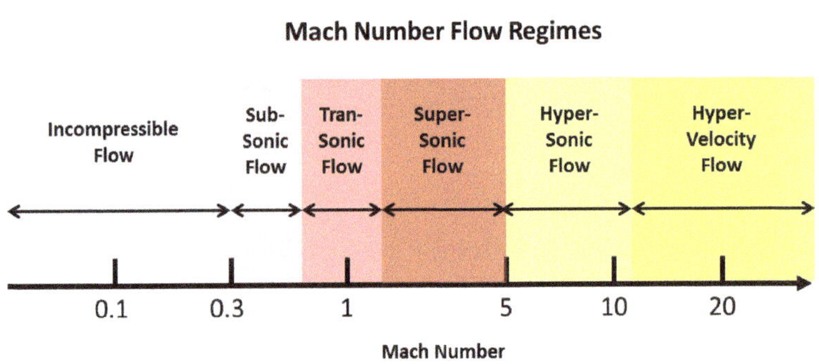

Mach number flow regimes spectrum

These flow regimes are not chosen arbitrarily, but rather arise naturally from the strong mathematical background that underlies compressible. At very slow flow speeds the speed of sound is so much faster that it is mathematically ignored, and the Mach number is irrelevant. Once the speed of the flow approaches the speed of sound, however, the Mach number becomes all-important, and shock waves begin to appear. Thus the transonic regime is described by a different (and much more difficult) mathematical treatment. In the supersonic regime the flow is dominated by wave motion at oblique angles similar to the Mach angle. Above about Mach 5, these wave angles grow so small that a different mathematical approach is required, defining the Hypersonic speed regime. Finally, at speeds comparable to that of planetary atmospheric entry from orbit, in the range of several km/s, the speed of sound is now comparatively so slow that it is once again mathematically ignored in the Hypervelocity regime.

As an object accelerates from subsonic toward supersonic speed in a gas, different types of wave phenomena occur. To illustrate these changes, the next figure shows a stationary point (M = 0) that emits symmetric sound waves. The speed of sound is the same in all directions in a uniform fluid, so these waves are simply concentric spheres. As the sound-generating point begins to accelerate, the sound waves "bunch up" in the direction of motion and "stretch out" in the opposite direction. When the point reaches sonic speed (M = 1), it travels at the same speed as the sound waves it creates. Therefore, an infinite number of these sound waves "pile up" ahead of the point, forming a Shock wave. Upon achieving supersonic flow, the particle is moving so fast that it continuously leaves its sound waves behind. When this occurs, the locus of these waves trailing behind the point creates an angle known as the Mach wave angle or Mach angle, μ:

$$\mu = \arcsin(a / V) = \arcsin(1 / M)$$

where a represents the speed of sound in the gas and V represents the velocity of the object. Although named for Austrian physicist Ernst Mach, these oblique waves were actually first discovered by Christian Doppler.

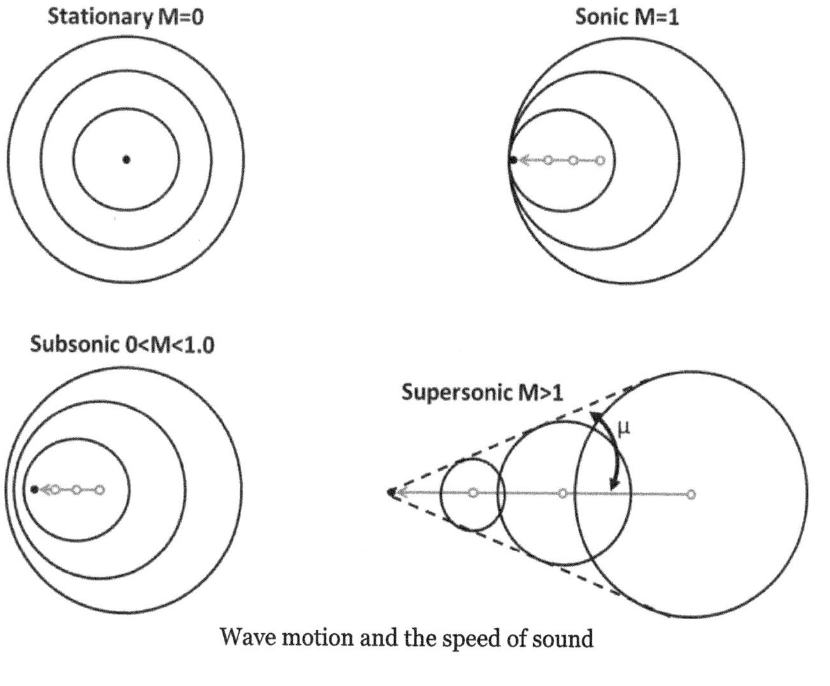

Wave motion and the speed of sound

One-dimensional Flow

One-dimensional (1-D) flow refers to the flow of gas through a duct or channel in which the flow parameters are assumed to change significantly along only one spatial dimension, namely, the duct length. In analysing the 1-D channel flow, a number of assumptions are made:

- Ratio of duct length to width (L/D) is ≤ about 5 (in order to neglect friction and heat transfer),
- Steady vs. Unsteady Flow,
- Flow is isentropic (i.e. a reversible adiabatic process),
- Ideal gas law (i.e. P=ρRT)

Converging-diverging Laval Nozzles

As the speed of a flow accelerates from the subsonic to the supersonic regime, the physics of nozzle and diffuser flows is altered. Using the conservation laws of fluid dynamics and thermodynamics, the following relationship for channel flow is developed (combined mass and momentum conservation):

$$dP(1-M^2) = \rho V^2 \left(\frac{dA}{A} \right),$$

where dP is the differential change in pressure, M is the Mach number, ρ is the density of the gas, V is the velocity of the flow, A is the area of the duct, and dA is the change in area of the duct. This equation states that, for subsonic flow, a converging duct (dA<0) increases the velocity of the flow and a diverging duct (dA>0) decreases velocity of the flow. For supersonic flow, the opposite occurs due to the change of sign of ($1-M^2$). A converging duct (dA<0) now decreases the velocity of the flow and a diverging duct (dA>0) increases the velocity of the flow. At Mach = 1, a special case occurs in which the duct area must be either a maximum or minimum. For practical purposes, only a minimum area can accelerate flows to Mach 1 and beyond.

	Subsonic M<1	Supersonic M>1
Converging Duct 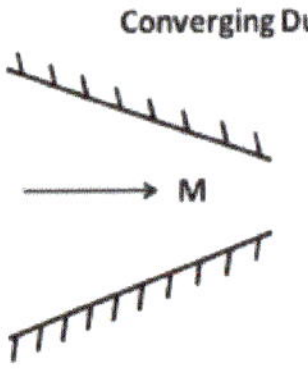	As P decreases, then *v* increases Subsonic nozzle	As P increases, then *v* decreases Supersonic diffuser
Diverging Duct 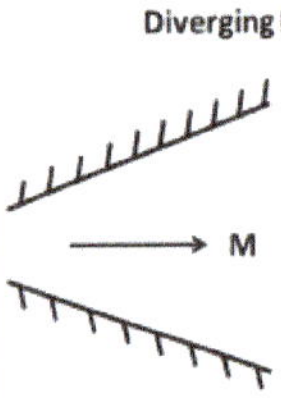	As P increases, then *v* decreases Subsonic diffuser	As P decreases, then *v* increases Supersonic nozzle

Table showing the reversal in the physics of nozzles and diffusers with changing Mach numbers

Therefore, to accelerate a flow to Mach 1, a nozzle must be designed to converge to a minimum cross-sectional area and then expand. This type of nozzle – the converging-diverging nozzle – is called a de Laval nozzle after Gustaf de Laval, who invented it. As subsonic flow enters the converging duct and the area decreases, the flow accelerates. Upon reaching the minimum area of the duct, also known as the throat of the nozzle, the flow can reach Mach 1. If the speed of the flow is to continue to increase, its density must decrease in order to obey conservation of mass. To achieve this decrease in density, the flow must expand, and to do so, the flow must pass through a diverging duct.

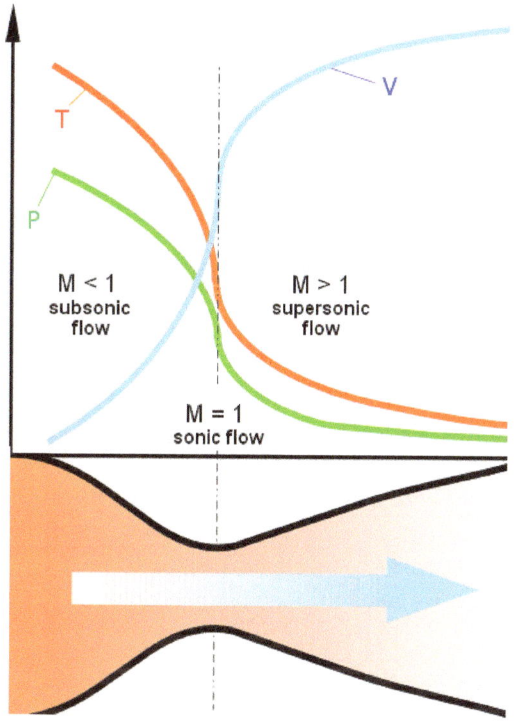

Nozzle de Laval diagram

Maximum Achievable Velocity of a Gas

Ultimately, because of the energy conservation law, a gas is limited to a certain maximum velocity based on its energy content. The maximum velocity, V_{max}, that a gas can attain is:

$$V_{max} = \sqrt{2c_p T_t}$$

where c_p is the specific heat of the gas and T_t is the stagnation temperature of the flow.

Isentropic Flow Mach Number Relationships

Using conservations laws and thermodynamics, a number of relationships of the form

$$\frac{\text{property}_1}{\text{property}_2} = f(M, \gamma)$$

can be obtained, where M is the Mach number and γ is the ratio of specific heats (1.4 for air).

Isentropic Flow Mach Number Relationships:	
Stagnation to Static Temperature	$\frac{T_t}{T} = \left(1 + \frac{\gamma - 1}{2} M^2\right)$
Stagnation to Static Pressure	$\frac{P_t}{P} = \left(1 + \frac{\gamma - 1}{2} M^2\right)^{\frac{\gamma}{\gamma - 1}}$
Mass Flow Rate	$\dot{m} = \frac{P_t}{\sqrt{RT_t}} A \sqrt{\gamma} M \left(1 + \frac{\gamma - 1}{2} M^2\right)^{\frac{\gamma + 1}{2 - 2\gamma}}$
Stagnation to Static Density	$\frac{\rho_t}{\rho} = \left(1 + \frac{\gamma - 1}{2} M^2\right)^{\frac{1}{\gamma - 1}}$
Arbitrary Area to Choked Area in Duct	$\frac{A}{A*} = \frac{1}{M} \left(\frac{\frac{\gamma + 1}{2}}{1 + \frac{\gamma - 1}{2} M^2}\right)^{\frac{\gamma + 1}{2 - 2\gamma}}$

Isentropic flow relationship table. Equations to relate the field properties in isentropic flow.

Achieving Supersonic Flow

As previously mentioned, in order for a flow to become supersonic, it must pass through a duct with a minimum area, or sonic throat. Additionally, an overall pressure ratio, P_b/P_t, of approximately 2 is needed to attain Mach 1. Once it has reached Mach 1, the flow at the throat is said to be "choked." Because changes downstream can only move upstream at sonic speed, the mass flow through the nozzle cannot be affected by changes in downstream conditions after the flow is choked.

Non-isentropic 1D Channel Flow of a Gas - normal Shock Waves

Normal shock waves are shock waves that are perpendicular to the local flow direction. These shock waves occur when pressure waves build up and coalesce into an extremely thin shockwave that converts useful energy into heat. The waves thus overtake and reinforce one another, forming a finite shock wave from an infinite series of infinitesimal sound waves. Because a loss of energy occurs over the thin shock wave, the shock is considered non-isentropic and enthalpy increases across the shock. When analysing a normal shock wave, one-dimensional, steady, and adiabatic (stagnation temperature does not change across the shock wave) flow of a perfect gas is assumed.

Rankine-Hugoniot Equations:	
Temperature	$\frac{T_2}{T_1} = \frac{\left(1 + \frac{\gamma - 1}{2} M_1^2\right)\left(\frac{2\gamma}{\gamma - 1} M_1^2 - 1\right)}{M_1^2 \left(\frac{2\gamma}{\gamma - 1} + \frac{\gamma - 1}{2}\right)}$
Static Pressure	$\frac{P_2}{P_1} = \frac{2\gamma M_1^2}{\gamma + 1} - \frac{\gamma - 1}{\gamma + 1}$
Density and Velocity	$\frac{\rho_2}{\rho_1} = \frac{v_2}{v_1} = \frac{(\gamma + 1) M_1^2}{(\gamma - 1) M_1^2 + 2}$

The Rankine-Hugoniot equations relate conditions before and after a normal shock wave.

Normal shock waves can occur in two reference frames: the standing normal shock and the moving shock. The flow before a normal shock wave must be supersonic, and the flow after a normal shock must be subsonic. The Rankine-Hugoniot equations are used to solve for the flow conditions.

Two-dimensional Flow

Although one-dimensional flow can be directly analysed, it is merely a specialized case of two-dimensional flow. It follows that one of the defining phenomena of one-dimensional flow, a normal shock, is likewise only a special case of a larger class of oblique shocks. Further, the name "normal" is with respect to geometry rather than frequency of occurrence. Oblique shocks are much more common in applications such as: aircraft inlet design, objects in supersonic flight, and (at a more fundamental level) supersonic nozzles and diffusers. Depending on the flow conditions, an oblique shock can either be attached to the flow or detached from the flow in the form of a bow shock.

Attached shock wave shown on a X-15 Model in a supersonic wind tunnel

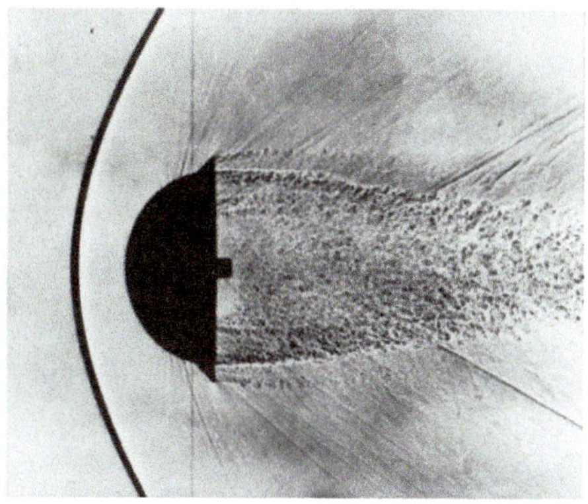

Bowshock example for a blunt body

Oblique Shock Waves

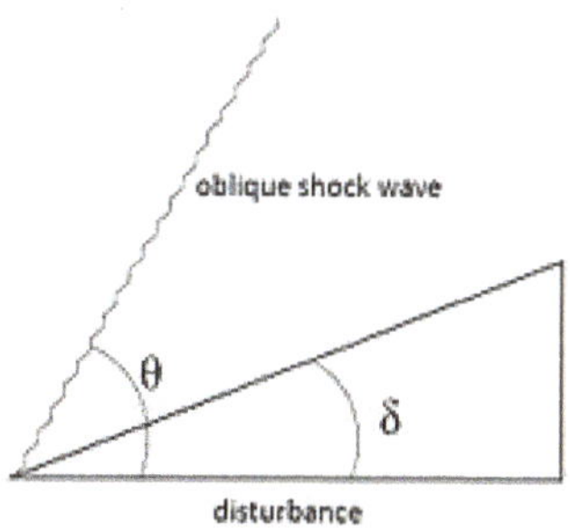

Diagram of obstruction

Oblique shock waves are similar to normal shock waves, but they occur at angles less than 90° with the direction of flow. When a disturbance is introduced to the flow at a nonzero angle (δ), the flow must respond to the changing boundary conditions. Thus an oblique shock is formed, resulting in a change in the direction of the flow.

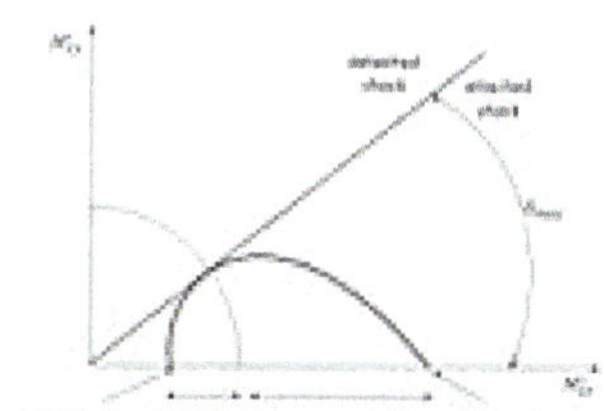

Shock polar diagram

Shock Polar Diagram

Based on the level of flow deflection (δ), oblique shocks are characterized as either strong or weak. Strong shocks are characterized by larger deflection and more entropy loss across the shock, with weak shocks as the opposite. In ordert to gain cursory insight into the differences in these shocks, a shock polar diagram can be used. With the static temperature after the shock, T*, known the speed of sound after the shock is defined as,

$$A^* = \sqrt{\gamma R T^*}$$

with R as the gas constant and γ as the specific heat ratio. The Mach number can be broken into Cartesian coordinates

$$M_{2x}^* = \frac{V_x}{a^*}$$

$$M_{2y}^* = \frac{V_y}{a^*}$$

with V_x and V_y as the x and y-components of the fluid velocity V. With the Mach number before the shock given, a locus of conditions can be specified. At some δ_{max} the flow transitions from a strong to weak oblique shock. With δ = 0°, a normal shock is produced at the limit of the strong oblique shock and the Mach wave is produced at the limit of the weak shock wave.

Oblique Shock Reflection

Due to the inclination of the shock, after an oblique shock is created, it can interact with a boundary in three different manners, two which are explained below.

Solid boundary

Incoming flow is first turned by angle δ with respect to the flow. This shockwave is reflected off the solid boundary, and the flow is turned by – δ to again be parallel with the boundary. It is important to note that each progressive shock wave is weaker and the wave angle is increased.

Irregular reflection

An irregular reflection is much like the case described above, with the caveat that δ is larger than the maximum allowable turning angle. Thus a detached shock is formed and a more complicated reflection occurs.

Prandtl-Meyer Fans

Prandtl-Meyer fans can be expressed as both compression and expansion fans. Prandtl-Meyer fans also cross a boundary layer (i.e. flowing and solid) which reacts in different changes as well. When a shock wave hits a solid surface the resulting fan returns as one from the opposite family while when one hits a free boundary the fan returns as a fan of opposite type.

Prandtl-Meyer Expansion Fans

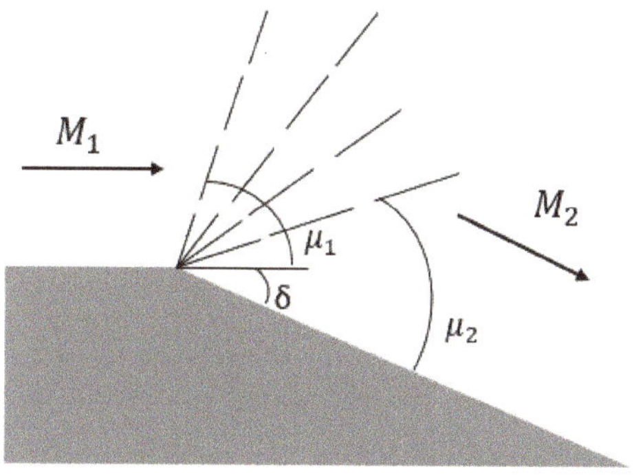

Prandtl-Meyer expansion fan diagram

To this point, the only flow phenomena that have been discussed are shock waves, which slow the flow and increase its entropy. It is possible to accelerate supersonic flow in what has been termed a Prandtl-Meyer expansion fan, after Ludwig Prandtl and Theodore Meyer. The mechanism for the expansion is shown in the figure below.

As opposed to the flow encountering an inclined obstruction and forming an oblique shock, the flow expands around a convex corner and forms an expansion fan through a series of isentropic Mach waves. The expansion "fan" is composed of Mach waves that span from the initial Mach angle to the final Mach angle. Flow can expand around either a sharp or rounded corner equally, as the increase in Mach number is proportional to only the convex angle of the passage (δ). The expansion corner that produces the Prandtl-Meyer fan can be sharp (as illustrated in the figure) or rounded. If the total turning angle is the same, then the P-M flow solution is also the same.

The Prandtl-Meyer expansion can be seen as the physical explanation of the operation of the Laval nozzle. The contour of the nozzle creates a smooth and continuous series of Prandtl-Meyer expansion waves.

Prandtl-Meyer Compression Fans

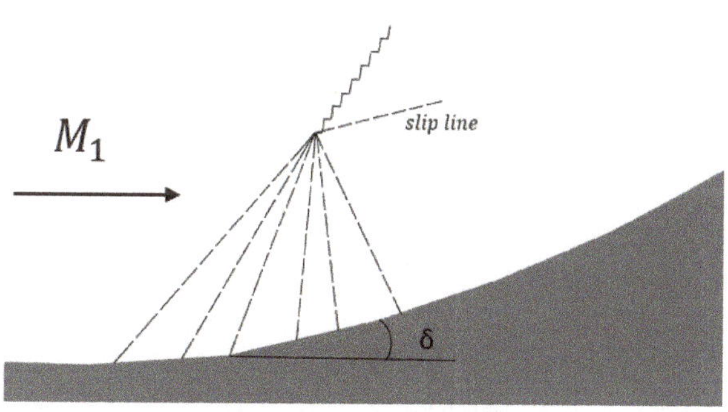

Basic PM compression diagram

A Prandtl-Meyer compression is the opposite phenomenon to a Prandtl-Meyer expansion. If the flow is gradually turned through an angle of δ, a compression fan can be formed. This fan is a series of Mach waves that eventually coalesce into an oblique shock. Because the flow is defined by an isentropic region (flow that travels through the fan) and an anisentropic region (flow that travels through the oblique shock), a slip line results between the two flow regions.

Applications

Supersonic Wind Tunnels

Supersonic wind tunnels are used for testing and research in supersonic flows, approximately over the Mach number range of 1.2 to 5. The operating principle behind the wind tunnel is that a large pressure difference is maintained upstream to downstream, driving the flow.

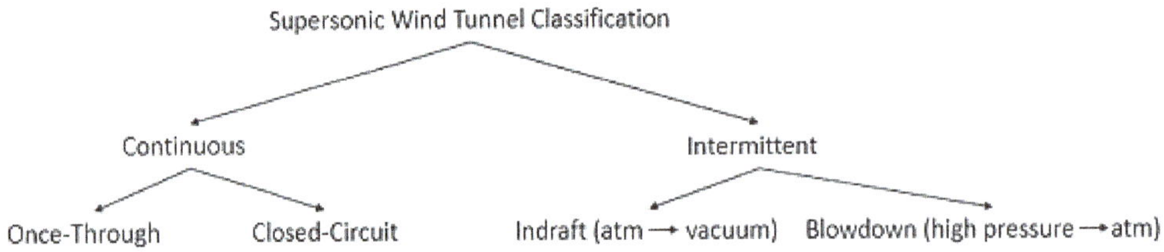

Supersonic wind tunnel classification list

Wind tunnels can be divided into two categories: continuous-operating and intermittent-operating wind tunnels. Continuous operating supersonic wind tunnels require an independent electrical power source that drastically increases with the size of the test section. Intermittent supersonic wind tunnels are less expensive in that they store electrical energy over an extended period of time, then discharge the energy over a series of brief tests. The difference between these two is analogous to the comparison between a battery and a capacitor.

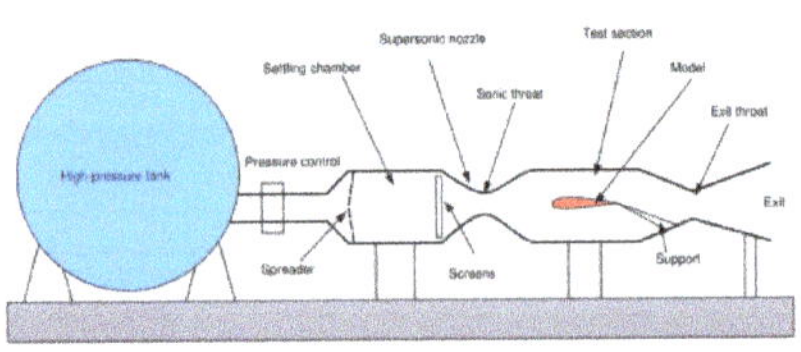

Blowdown supersonic wind tunnel schematic

Langley indraft supersonic wind tunnel vacuum sphere

Blowdown type supersonic wind tunnels offer high Reynolds number, a small storage tank, and readily available dry air. However, they cause a high pressure hazard, result in difficulty holding a constant stagnation pressure, and are noisy during operation.

Indraft supersonic wind tunnels are not associated with a pressure hazard, allow a constant stagnation pressure, and are relatively quiet. Unfortunately, they have a limited range for the Reynolds number of the flow and require a large vacuum tank. There is no dispute that knowledge is gained through research and testing in supersonic wind tunnels; however, the facilities often require vast amounts of power to maintain the large pressure ratios needed for testing conditions. For example, Arnold Engineering Development Complex has the largest supersonic wind tunnel in the world and requires the power required to light a small city for operation. For this reason, large wind tunnels are becoming less common at universities.

Supersonic Aircraft Inlets

Perhaps the most common application for oblique shocks is in high-speed aircraft inlets. The purpose of the inlet is to slow incoming supersonic flow to the subsonic regime before it enters the turbojet engine, with the caveat of minimizing losses across the shock. Knowledge of normal and oblique shocks suggests that this be accomplished with a series of weakening oblique shocks followed by a very weak normal shock, usually less than M = 1.4. This may sound relatively straightforward, but there is one rather large issue to be dealt with when designing a supersonic aircraft inlet: acceleration. Between taking off, manoeuvring, and cruising, an aircraft travels at a range of Mach numbers. In order to ensure efficient flight, the aircraft intake must be capable of variable geometry. If it is not, the shock waves will not reflect properly through the inlet and negatively affect performance.

McDonnell Douglas F-15C

Concorde

Lockheed SR-71 Blackbird

Although variable geometry is a universally recognized approach to improve aircraft efficiency and performance over a range of Mach numbers, there is no one method to achieve variable geometry. The F-15 Eagle employs wedge inlets with adjustable flaps to control the flow. For subsonic flow, the flaps are completely closed and for supersonic flow, the flaps are open. The Concorde employed an external-compression inlet, using a series of oblique shocks followed by a normal shock to slow the flow sufficiently for the turbojet engine. Perhaps the most recognizable supersonic aircraft, the SR-71, used a hydraulically actuated cone to reduce the speed of the supersonic flow through the aircraft inlet.

Natural Gas Pipeline

Constructing natural gas line in winter, Finland

Natural gas pipelines are used to transport natural gas from extraction sites to refinement or chemical processing facilities. In the United States there are more than 210 natural gas pipeline systems with more than 305,000 miles (491,000 km) of intrastate transmission pipelines. Two compressible flow phenomena characterize the flow through these pipelines: friction (Fanno flow and Rayleigh flow) and heat transfer. Natural gas pipelines are buried in the ground at a constant temperature of 15 °C. However, the friction generated by the flow offsets the heat loss to the Earth, thus resulting in an isothermal flow.

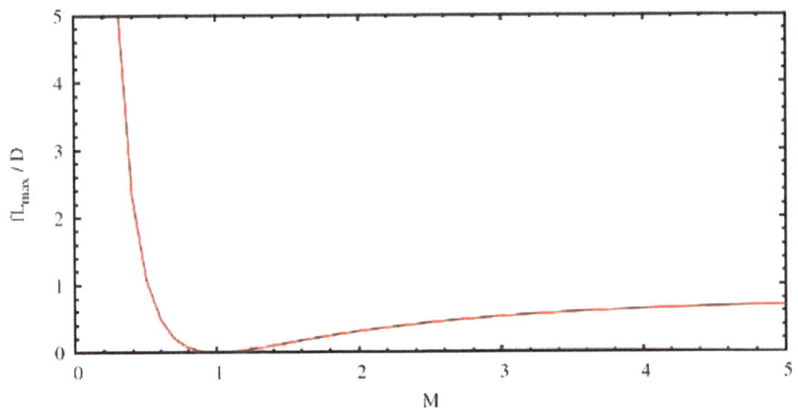

Friction factor as a function of Mach number in Fanno flow

The relationship between fL_{max}/D and Mach number for Fanno flow suggests that only subsonic flow can be used in the long pipes used to transport natural gas (even these pipes must be broken into shorter segments with compressor stations at the discontinuities in the pipeline). Additionally using conservation, an equation can be derived to describe the flow.

$$\frac{dM}{M} = \frac{fdx}{D} \frac{\gamma M^2}{2 - 2\gamma M^2}$$

This equation describes flow that chokes at $M^* = 0.87$ for natural gas $\gamma = 1.32$; however choking requires an infinite heat flux. Therefore, a combination of intuition and mathematics explains why it is most economically feasible that subsonic natural gas is pumped through long sections of pipe to reach its intended destination.

Shock Absorber

Miniature oil filled Coilover shock components for scale cars.

A shock absorber (in reality, a shock "damper") is a mechanical or hydraulic device designed to absorb and damp shock impulses. It does this by converting the kinetic energy of the shock into another form of energy (typically heat) which is then dissipated. Most shock absorbers are a form of dashpot.

Description

Pneumatic and hydraulic shock absorbers are used in conjunction with cushions and springs. An automobile shock absorber contains spring-loaded check valves and orifices to control the flow of oil through an internal piston.

One design consideration, when designing or choosing a shock absorber, is where that energy will go. In most shock absorbers, energy is converted to heat inside the viscous fluid. In hydraulic cylinders, the hydraulic fluid heats up, while in air cylinders, the hot air is usually exhausted to the atmosphere. In other types of shock absorbers, such as electromagnetic types, the dissipated energy can be stored and used later. In general terms, shock absorbers help cushion vehicles on uneven roads.

Vehicle Suspension

In a vehicle, shock absorbers reduce the effect of traveling over rough ground, leading to improved ride quality and vehicle handling. While shock absorbers serve the purpose of limiting excessive suspension movement, their intended sole purpose is to damp spring oscillations. Shock absorbers use valving of oil and gasses to absorb excess energy from the springs. Spring rates are chosen by the manufacturer based on the weight of the vehicle, loaded and unloaded. Some people use shocks to modify spring rates but this is not the correct use. Along with hysteresis in the tire itself, they damp the energy stored in the motion of the unsprung weight up and down. Effective wheel bounce damping may require tuning shocks to an optimal resistance.

Spring-based shock absorbers commonly use coil springs or leaf springs, though torsion bars are used in torsional shocks as well. Ideal springs alone, however, are not shock absorbers, as springs only store and do not dissipate or absorb energy. Vehicles typically employ both hydraulic shock absorbers and springs or torsion bars. In this combination, "shock absorber" refers specifically to the hydraulic piston that absorbs and dissipates vibration. Now, composite suspension system are used mainly in 2 wheelers and also leaf spring are made up of composite material in 4 wheelers.

Early History

In common with carriages and railway locomotives, most early motor vehicles used leaf springs. One of the features of these springs was that the friction between the leaves offered a degree of damping, and in a 1912 review of vehicle suspension the lack of this characteristic in helical springs was the reason it was "impossible" to use them as main springs. However the amount of damping provided by leaf spring friction was limited and variable according to the conditions of the springs, and whether wet or dry. It also operated in both directions. Motorcycle front suspension adopted coil sprung Druid forks from about 1906, and similar designs later added rotary friction dampers, which damped both ways - but they were adjustable (e.g. 1924 Webb forks). These friction disk shock absorbers were also fitted to many cars.

One of the problems with motor cars was the large variation in sprung weight between lightly loaded and fully loaded, especially for the rear springs. When heavily loaded the springs could bottom out, and apart from fitting rubber 'bump stops', there were attempts to use heavy main springs with auxiliary springs to smooth the ride when lightly loaded, which were often called 'shock absorbers'. Realising that the spring and vehicle combination bounced with a characteristic frequency, these auxiliary springs were designed with a different period, but were not a solution to the problem that the spring rebound after striking a bump could throw you out of your seat. What was called for was damping that operated on the rebound.

Although C.L. Horock came up with a design in 1901 that had hydraulic damping, it worked in one direction only. It does not seem to have gone into production right away, whereas mechanical dampers such as the Gabriel Snubber started being fitted in the late 1900s (also the similar Stromberg Anti-Shox). These used a belt coiled inside a device such that it freely wound in under the action of a coiled spring, but met friction when drawn out. Gabriel Snubbers were fitted to an 11.9HP Arrol-Johnston car which broke the 6 hour Class B record at Brooklands in late 1912, and the Automotor journal noted that this snubber might have a great future for racing due to its light weight and easy fitment.

One of the earliest hydraulic dampers to go into production was the Telesco Shock Absorber, exhibited at the 1912 Olympia Motor Show and marketed by Polyrhoe Carburettors Ltd. This contained a spring inside the telescopic unit like the pure spring type 'shock absorbers' mentioned above, but also oil and an internal valve so that the oil damped in the rebound direction. The Telesco unit was fitted at the rear end of the leaf spring, in place of the rear spring to chassis mount, so that it formed part of the springing system, albeit a hydraulically damped part. This layout was presumably selected as it was easy to apply to existing vehicles, but it meant the hydraulic damping was not applied to the action of the main leaf spring, but only to the action of the auxiliary spring in the unit itself.

The first production hydraulic dampers to act on the main leaf spring movement were probably those based on an original concept by Maurice Houdaille patented in 1908 and 1909. These used a lever arm which moved hydraulically damped vanes inside the unit. The main advantage over the friction disk dampers was that it would resist sudden movement but allow slow movement, whereas the rotary friction dampers tended to stick and then offer the same resistance regardless of speed of movement. There appears to have been little progress on commercialising the lever arm shock aborbers until after World War I, after which they came into widespread use, for example as standard equipment on the 1927 Ford Model A.

Types of Vehicle Shock Absorbers

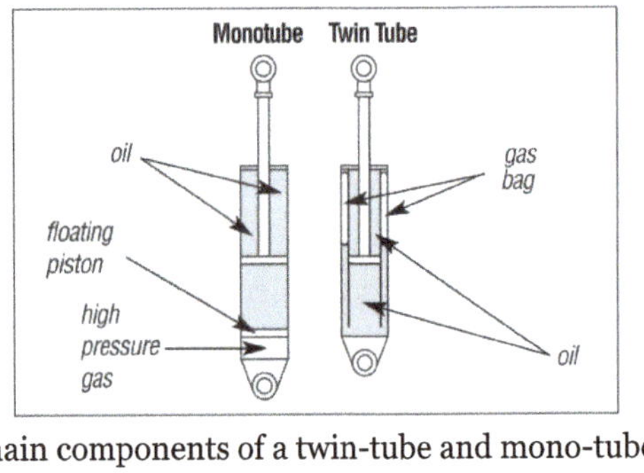

Diagram of the main components of a twin-tube and mono-tube shock absorber

Most vehicular shock absorbers are either twin-tube or mono-tube types with some variations on these themes.

Twin-Tube

Basic Twin-tube

Also known as a "two-tube" shock absorber, this device consists of two nested cylindrical tubes, an inner tube that is called the "working tube" or the "pressure tube", and an outer tube called the "reserve tube". At the bottom of the device on the inside is a compression valve or base valve. When the piston is forced up or down by bumps in the road, hydraulic fluid moves between different chambers via small holes or "orifices" in the piston and via the valve, converting the "shock" energy into heat which must then be dissipated.

Twin-tube Gas Charged

Variously known as a "gas cell two-tube" or similarly-named design, this variation represented a significant advancement over the basic twin-tube form. Its overall structure is very similar to the twin-tube, but a low-pressure charge of nitrogen gas is added to the reserve tube. The result of this alteration is a dramatic reduction in "foaming" or "aeration", the undesirable outcome of a twin-tube overheating and failing which presents as foaming hydraulic fluid dripping out of the assembly. Twin-tube gas charged shock absorbers represent the vast majority of original modern vehicle suspensions installations.

Position Sensitive Damping

Often abbreviated simply as "PSD", this design is another evolution of the twin-tube shock. In a PSD shock absorber, which still consists of two nested tubes and still contains nitrogen gas, a set of grooves has been added to the pressure tube. These grooves allow the piston to move relatively freely in the middle range of travel (i.e., the most common street or highway use, called by engineers the "comfort zone") and to move with significantly less freedom in response to shifts to more irregular surfaces when upward and downward movement of the piston starts to occur with greater intensity (i.e., on bumpy sections of roads— the stiffening gives the driver greater control of movement over the vehicle so its range on either side of the comfort zone is called the "control zone"). This advance allowed car designers to make a shock absorber tailored to specific makes and models of vehicles and to take into account a given vehicle's size and weight, its maneuverability, its horsepower, etc. in creating a correspondingly effective shock.

Acceleration Sensitive Damping

The next phase in shock absorber evolution was the development of a shock absorber that could sense and respond to not just situational changes from "bumpy" to "smooth" but to individual bumps in the road in a near instantaneous reaction. This was achieved through a change in the design of the compression valve, and has been termed "acceleration sensitive damping" or "ASD". Not only does this result in a complete disappearance of the "comfort vs. control" tradeoff, it also reduced pitch during vehicle braking and roll during turns. However, ASD shocks are usually only available as aftermarket changes to a vehicle and are only available from a limited number of manufacturers.

Coilover

Coilover shock absorbers are usually a kind of twin-tube gas charged shock absorber around which has been mounted a large metal coil. Though common on motorcycle and scooter rear suspensions, coilover shocks are uncommon in original equipment designs for vehicles, though they have become widely available as aftermarket add-ons. Coilover shocks for cars have been considered specialty items for high performance and racing applications where they allow for significant reductions in overall vehicle height, and though high-quality aftermarket options with wide sturdy springs may provide improvements in vehicle performance, there is dispute over whether or not most aftermarket coilover shocks confer any material benefits to most drivers and may in fact reduce performance over original equipment installations.

Mono-tube

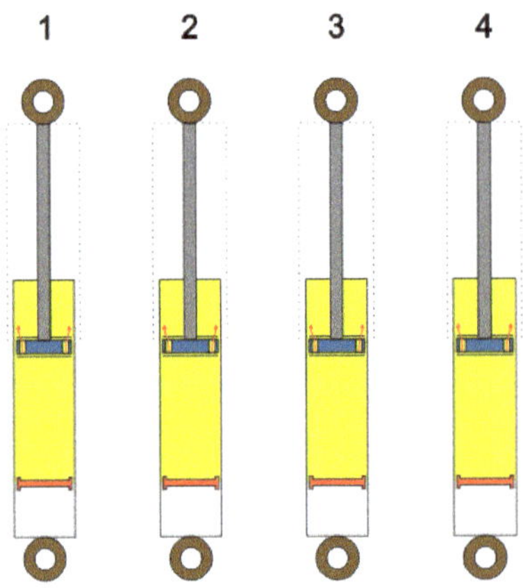

Hydraulic shock absorber monotube in different operational situations:

1) Drive slow or adjustments open

2) Like "1", but extension immediately after the compression

3) Drive fast adjustments or closed, you can see the bubles of depression, which can lead to the phenomenon of cavitation

4) Like "3", but the extension immediately after the compressionNote: The volume change caused by the stem is considered.

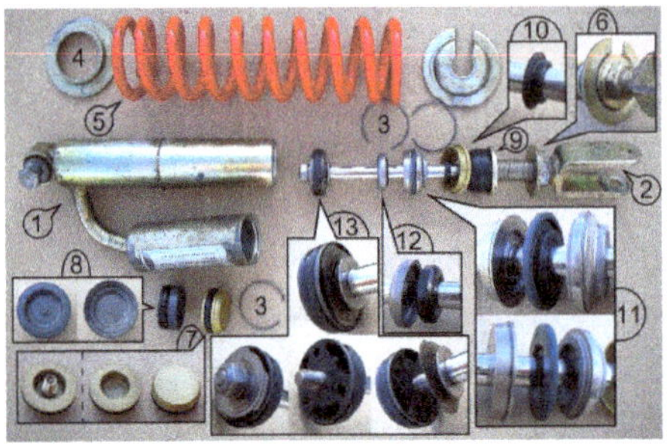

Absorber with gas tank connected rigidly, compared to most shock absorbers. It uses a diaphragm instead of a membrane, and does not contain a control valve for expansion of the pneumatic chamber.
Description:

1) Sheath and gas tank

2) Stem

3) Snap rings

4) Plate bearing spring

5) Spring

6) End cap and preload adjustment

7) Cap gas, present in versions both with or without gas valve (inverted profile)

8) Mobile diaphragm

9) Pad switch (compression)

10) Wiper

11) Oil seal assembly, and shock seal

12) Negative buffer pad or limit switch (extension)

13) Piston with sliding blades and seal

The principal design alternative to the twin-tube form has been the mono-tube shock absorber which was considered a revolutionary advancement when it appeared in the 1950s. As its name implies, the mono-tube shock, which is also a gas-pressurized shock and also comes in a coilover format, consists of only one tube, the pressure tube, though it has two pistons. These pistons are called the working piston and the dividing or floating piston, and they move in relative synchrony inside the pressure tube in response to changes in road smoothness. The two pistons also completely separate the shock's fluid and gas components. The mono-tube shock absorber is consistently a much longer overall design than the twin-tubes, making it difficult to mount in passenger cars designed for twin-tube shocks. However, unlike the twin-tubes, the mono-tube shock can be mounted either way— it does not have any directionality. It also does not have a compression valve, whose role has been taken up by the dividing piston, and although it contains nitrogen gas, the gas in a mono-tube shock is under *high* pressure (260-360 p.s.i. or so) which can actually help it to support some of the vehicle's weight, something which no other shock absorber is designed to do.

Mercedes became the first auto manufacturer to install mono-tube shocks as standard equipment on some of their cars starting in 1958, manufactured by Bilstein. Because the design was patented, no other manufacturer could use it until 1971 when the patent expired.

Theoretical Approaches

There are several commonly used principles behind shock absorption:

- Hysteresis of structural material, for example the compression of rubber disks, stretching of rubber bands and cords, bending of steel springs, or twisting of torsion bars. Hysteresis is the tendency for otherwise elastic materials to rebound with less force than was required to deform them. Simple vehicles with no separate shock absorbers are damped, to some extent, by the hysteresis of their springs and frames.

- Dry friction as used in wheel brakes, by using disks (classically made of leather) at the pivot of a lever, with friction forced by springs. Used in early automobiles such as the Ford Model T, up through some British cars of the 1940s. Although now considered obsolete, an advantage of this system is its mechanical simplicity; the degree of damping can be easily adjusted by tightening or loosening the screw clamping the disks, and it can be easily rebuilt with simple hand tools. A disadvantage is that the damping force tends not to increase with the speed of the vertical motion.

- Solid state, tapered chain shock absorbers, using one or more tapered, axial alignment(s) of granular spheres, typically made of metals such as nitinol, in a casing.

- Fluid friction, for example the flow of fluid through a narrow orifice (hydraulics), constitutes the vast majority of automotive shock absorbers. This design first appeared on Mors racing cars in 1902. One advantage of this type is, by using special internal valving, the absorber may be made relatively soft to compression (allowing a soft response to a bump) and relatively stiff to extension, controlling "rebound", which is the vehicle response to energy stored in the springs; similarly, a series of valves controlled by springs can change the degree of stiffness according to the velocity of the impact or rebound. Specialized shock absorbers for racing purposes may allow the front end of a dragster to rise with minimal resistance under acceleration, then strongly resist letting it settle, thereby maintaining a desirable rearward weight distribution for enhanced traction.

- Compression of a gas, for example pneumatic shock absorbers, which can act like springs as the air pressure is building to resist the force on it. Enclosed gas is compressible, so equipment is less subject to shock damage. This concept was first applied in series production on Citroën cars in 1954. Today, many shock absorbers are pressurized with compressed nitrogen, to reduce the tendency for the oil to cavitate under heavy use. This causes foaming which temporarily reduces the damping ability of the unit. In very heavy duty units used for racing or off-road use, there may even be a secondary cylinder connected to the shock absorber to act as a reservoir for the oil and pressurized gas. In aircraft landing gear, air shock absorbers may be combined with hydraulic damping to reduce bounce. Such struts are called *oleo struts* (combining oil and air).

- Inertial resistance to acceleration, for example prior to 1966 , the Citroën 2CV had shock absorbers that damp wheel bounce with no external moving parts. These consisted of a spring-mounted 3.5 kg (7.75 lb) iron weight inside a vertical cylinder and are similar to, yet much smaller than versions of the tuned mass dampers used on tall buildings.

- Composite hydropneumatic suspension combines many suspension elements in a single device: spring action, shock absorption, ride-height control, and self leveling suspension. This combines the advantages of gas compressibility and the ability of hydraulic machinery to apply force multiplication.

- Conventional shock absorbers can be combined with air suspension springs - an alternate way to achieve ride-height control, and self leveling suspension.

- In an electrorheological fluid damper, an electric field changes the viscosity of the oil. This principle allows semi-active damper applications in automotive and various industries.
- Magnetic field variation: a magneto rheological damper changes its fluid characteristics through an electromagnet.
- The effect of a shock absorber at high (sound) frequencies is usually limited by using a compressible gas as the working fluid or mounting it with rubber bushings.

Special Features

- Some shock absorbers allow tuning of the ride via control of the valve by a manual adjustment provided at the shock absorber.
- In more expensive vehicles the valves may be remotely adjustable, offering the driver control of the ride at will while the vehicle is operated.
- Additional control can be provided by dynamic valve control via computer in response to sensors, giving both a smooth ride and a firm suspension when needed, allowing ride height adjustment or even ride height control.
- Ride height control is especially desirable in highway vehicles intended for occasional rough road use, as a means of improving handling and reducing aerodynamic drag by lowering the vehicle when operating on improved high speed roads.

References

- Goldman, Eric, "Wikipedia's Labor Squeeze and its Consequences", Journal on Telecommunications and High Technology Law, 8
- Carley, Larry (February 2008), "Monotube shocks-- don't absorb shocks, but..." (PDF), Brake and front end magazine, retrieved 1 January 2014
- LeVeque, Randall (2002). Finite Volume Methods for Hyperbolic Problems. Cambridge University Press. ISBN 978-0-521-00924-9
- Noveck, Beth Simone (March 2007), "Wikipedia and the Future of Legal Education", Journal of Legal Education, 57 (1)
- Toro, E. F. (1999). Riemann Solvers and Numerical Methods for Fluid Dynamics. Springer-Verlag. ISBN 978-3-540-65966-2
- Bill Venners (October 20, 2003). "Exploring with Wiki: A Conversation with Ward Cunningham, Part I". artima developer. Retrieved December 12, 2014
- Myers, Ken S. (2008), "Wikimmunity: Fitting the Communications Decency Act to Wikipedia", Harvard Journal of Law and Technology, The Berkman Center for Internet and Society, 20: 163, SSRN 916529
- Rich, Ben R.; Janos, Leo (1994). Skunk works: a personal memoir of my years at Lockheed. Warner Books. p. 218. ISBN 0751515035
- "List of largest (Media)wikis". S23-Wiki. April 3, 2008. Archived from the original on August 25, 2014. Retrieved December 12, 2014
- Veeser, L. R.; Solem, J. C. (1978). "Studies of Laser-driven shock waves in aluminum". Physical Review Letters. 40 (21): 1391. Bibcode:1978PhRvL..40.1391V. doi:10.1103/PhysRevLett.40.1391

- Johnson, Clarence L.; Smith, Maggie (1985). Kelly: more than my share of it all. Washington, D.C.: Smithsonian Institution Press. p. 141. ISBN 0874744911
- Cubric, Marija (2007). "Analysis of the use of Wiki-based collaborations in enhancing student learning". University of Hertfordshire. Retrieved April 25, 2011
- Anderson, John D. Jr. (2003) [1982]. Modern Compressible Flow (3rd ed.). McGraw-Hill Science/Engineering/Math. ISBN 0-07-242443-5
- Rankine, W. J. M. (1870). "On the thermodynamic theory of waves of finite longitudinal disturbances". Philosophical Transactions of the Royal Society of London. 160: 277–288. doi:10.1098/rstl.1870.0015
- Kurganov, V.A. (3 February 2011), Adiabatic Wall Temperature, Thermopedia, doi:10.1615/AtoZ.a.adiabatic_wall_temperature, retrieved 2015-10-03
- Oosthuizen, Patrick H.; Carscallen, W. E. (2013) [1997]. Introduction to Compressible Flow (2nd ed.). CRC Press. ISBN 1439877912

3

Pressure Measurement: Devices and Methods

The main instrument used in measuring pressure is called a vacuum gauge. It is used to measure pressure which is lower than the surrounding atmospheric pressure. Other scientific instruments explained in the following chapter are pressure sensor, McLeod gauge, Pirani gauge, hot-filament ionization gauge and thermocouple gauge. The topics discussed in the chapter are of great importance to broaden the existing knowledge on aerodynamics.

Pressure Measurement

Example of the widely used Bourdon pressure gauge

Many techniques have been developed for the measurement of pressure and vacuum. Instruments used to measure and display pressure in an integral unit are called pressure gauges or vacuum gauges. A manometer is a good example as it uses a column of liquid to both measure and indicate pressure. Likewise the widely used Bourdon gauge is a mechanical device which both measures and indicates, and is probably the best known type of gauge.

A vacuum gauge is an absolute pressure gauge used to measure the pressures lower than the ambient atmospheric pressure.

Other methods of pressure measurement involve sensors which can transmit the pressure reading to a remote indicator or control system (telemetry).

Absolute, Gauge and Differential Pressures — Zero Reference

Everyday pressure measurements, such as for vehicle tire pressure, are usually made relative to ambient air pressure. In other cases measurements are made relative to a vacuum or to some other specific reference. When distinguishing between these zero references, the following terms are used:

- Absolute pressure is zero-referenced against a perfect vacuum, using an absolute scale, so it is equal to gauge pressure plus atmospheric pressure.
- Gauge pressure is zero-referenced against ambient air pressure, so it is equal to absolute pressure minus atmospheric pressure. Negative signs are usually omitted. To distinguish a negative pressure, the value may be appended with the word "vacuum" or the gauge may be labeled a "vacuum gauge." These are further divided into two subcategories: high and low vacuum (and sometimes ultra-high vacuum). The applicable pressure ranges of many of the techniques used to measure vacuums have an overlap. Hence, by combining several different types of gauge, it is possible to measure system pressure continuously from 10 mbar down to 10^{-11} mbar.
- Differential pressure is the difference in pressure between two points.

The zero reference in use is usually implied by context, and these words are added only when clarification is needed. Tire pressure and blood pressure are gauge pressures by convention, while atmospheric pressures, deep vacuum pressures, and altimeter pressures must be absolute.

For most working fluids where a fluid exists in a closed system, gauge pressure measurement prevails. Pressure instruments connected to the system will indicate pressures relative to the current atmospheric pressure. The situation changes when extreme vacuum pressures are measured; absolute pressures are typically used instead.

Differential pressures are commonly used in industrial process systems. Differential pressure gauges have two inlet ports, each connected to one of the volumes whose pressure is to be monitored. In effect, such a gauge performs the mathematical operation of subtraction through mechanical means, obviating the need for an operator or control system to watch two separate gauges and determine the difference in readings.

Moderate vacuum pressure readings can be ambiguous without the proper context, as they may represent absolute pressure or gauge pressure without a negative sign. Thus a vacuum of 26 inHg gauge is equivalent to an absolute pressure of 30 inHg (typical atmospheric pressure) – 26 inHg = 4 inHg.

Atmospheric pressure is typically about 100 kPa at sea level, but is variable with altitude and weather. If the absolute pressure of a fluid stays constant, the gauge pressure of the same fluid will vary as atmospheric pressure changes. For example, when a car drives up a mountain, the (gauge) tire pressure goes up because atmospheric pressure goes down. The absolute pressure in the tire is essentially unchanged.

Using atmospheric pressure as reference is usually signified by a *g* for gauge after the pressure unit, e.g. 70 psig, which means that the pressure measured is the total pressure minus atmospheric pressure. There are two types of gauge reference pressure: vented gauge (vg) and sealed gauge (sg).

A vented gauge pressure transmitter for example allows the outside air pressure to be exposed to the negative side of the pressure sensing diaphragm, via a vented cable or a hole on the side of the device, so that it always measures the pressure referred to ambient barometric pressure. Thus a vented gauge reference pressure sensor should always read zero pressure when the process pressure connection is held open to the air.

A sealed gauge reference is very similar except that atmospheric pressure is sealed on the negative side of the diaphragm. This is usually adopted on high pressure ranges such as hydraulics where atmospheric pressure changes will have a negligible effect on the accuracy of the reading, so venting is not necessary. This also allows some manufacturers to provide secondary pressure containment as an extra precaution for pressure equipment safety if the burst pressure of the primary pressure sensing diaphragm is exceeded.

There is another way of creating a sealed gauge reference and this is to seal a high vacuum on the reverse side of the sensing diaphragm. Then the output signal is offset so the pressure sensor reads close to zero when measuring atmospheric pressure.

A sealed gauge reference pressure transducer will never read exactly zero because atmospheric pressure is always changing and the reference in this case is fixed at 1 bar.

To produce an absolute pressure sensor the manufacturer will seal a high vacuum behind the sensing diaphragm. If the process pressure connection of an absolute pressure transmitter is open to the air, it will read the actual barometric pressure.

Units

Pressure units						
v · t · e	**Pascal**	**Bar**	**Technical atmosphere**	**Standard atmosphere**	**Torr**	**Pounds per square inch**
	(Pa)	**(bar)**	**(at)**	**(atm)**	**(Torr)**	**(psi)**
1 Pa	≡ 1 N/m²	10^{-5}	1.0197×10^{-5}	9.8692×10^{-6}	7.5006×10^{-3}	1.450377×10^{-4}
1 bar	10^5	≡ 100 kPa ≡ 10^6 dyn/cm²	1.0197	0.98692	750.06	14.50377
1 at	9.80665×10^4	0.980665	≡ 1 kp/cm²	0.9678411	735.5592	14.22334
1 atm	1.01325×10^5	1.01325	1.0332	1	≡ 760	14.69595
1 Torr	133.3224	1.333224×10^{-3}	1.359551×10^{-3}	≡ 1/760 ≈ 1.315789×10^{-3}	≡ 1 Torr ≈ 1 mmHg	1.933678×10^{-2}
1 psi	6.8948×10^3	6.8948×10^{-2}	7.03069×10^{-2}	6.8046×10^{-2}	51.71493	≡ 1 lbf /in²

The SI unit for pressure is the pascal (Pa), equal to one newton per square metre ($N \cdot m^{-2}$ or $kg \cdot m^{-1} \cdot s^{-2}$). This special name for the unit was added in 1971; before that, pressure in SI was expressed in units such as $N \cdot m^{-2}$. When indicated, the zero reference is stated in parenthesis following the unit, for example 101 kPa (abs). The pound per square inch (psi) is still in widespread use in the US and Canada, for measuring, for instance, tire pressure. A letter is often appended to the psi unit to indicate the measurement's zero reference; psia for absolute, psig for gauge, psid for differential, although this practice is discouraged by the NIST.

Because pressure was once commonly measured by its ability to displace a column of liquid in a manometer, pressures are often expressed as a depth of a particular fluid (*e.g.*, inches of water). Manometric measurement is the subject of pressure head calculations. The most common choices for a manometer's fluid are mercury (Hg) and water; water is nontoxic and readily available, while mercury's density allows for a shorter column (and so a smaller manometer) to measure a given pressure. The abbreviation "W.C." or the words "water column" are often printed on gauges and measurements that use water for the manometer.

Fluid density and local gravity can vary from one reading to another depending on local factors, so the height of a fluid column does not define pressure precisely. So measurements in "millimetres of mercury" or "inches of mercury" can be converted to SI units as long as attention is paid to the local factors of fluid density and gravity. Temperature fluctuations change the value of fluid density, while location can affect gravity.

Although no longer preferred, these manometric units are still encountered in many fields. Blood pressure is measured in millimetres of mercury in most of the world, and lung pressures in centimeters of water are still common, as in settings for CPAP machines. Natural gas pipeline pressures are measured in inches of water, expressed as "inches W.C." Scuba divers often use a manometric rule of thumb: the pressure exerted by ten meters depth of sea water ("10 msw") is approximately equal to one atmosphere. In vacuum systems, the units torr (millimeter of mercury), micron (micrometer of mercury), and inch of mercury (inHg) are most commonly used. Torr and micron usually indicates an absolute pressure, while inHg usually indicates a gauge pressure.

Atmospheric pressures are usually stated using hectopascal (hPa), kilopascal (kPa), millibar (mbar) or atmospheres (atm). In American and Canadian engineering, stress is often measured in kip. Note that stress is not a true pressure since it is not scalar. In the cgs system the unit of pressure was the barye (ba), equal to 1 $dyn \cdot cm^{-2}$. In the mts system, the unit of pressure was the pieze, equal to 1 sthene per square metre.

Many other hybrid units are used such as $mmHg/cm^2$ or grams-force/cm^2 (sometimes as [[kg/cm^2]] without properly identifying the force units). Using the names kilogram, gram, kilogram-force, or gram-force (or their symbols) as a unit of force is prohibited in SI; the unit of force in SI is the newton (N).

Static and Dynamic Pressure

Static pressure is uniform in all directions, so pressure measurements are independent of direction in an immovable (static) fluid. Flow, however, applies additional pressure on surfaces perpendicular to the flow direction, while having little impact on surfaces parallel to the flow direction. This directional component of pressure in a moving (dynamic) fluid is called dynamic pressure. An instrument facing the flow direction measures the sum of the static and dynamic pressures; this measurement is called the total pressure or stagnation pressure. Since dynamic pressure is referenced to static pressure, it is neither gauge nor absolute; it is a differential pressure.

While static gauge pressure is of primary importance to determining net loads on pipe walls, dynamic pressure is used to measure flow rates and airspeed. Dynamic pressure can be measured by taking the differential pressure between instruments parallel and perpendicular to the flow. Pitot-static tubes, for example perform this measurement on airplanes to determine airspeed. The presence of the measuring instrument inevitably acts to divert flow and create turbulence, so its shape is critical to accuracy and the calibration curves are often non-linear.

Applications

- Altimeter
- Barometer
- MAP sensor
- Pitot tube
- Sphygmomanometer

Instruments

Many instruments have been invented to measure pressure, with different advantages and disadvantages. Pressure range, sensitivity, dynamic response and cost all vary by several orders of magnitude from one instrument design to the next. The oldest type is the liquid column (a vertical tube filled with mercury) manometer invented by Evangelista Torricelli in 1643. The U-Tube was invented by Christiaan Huygens in 1661.

Hydrostatic

Hydrostatic gauges (such as the mercury column manometer) compare pressure to the hydrostatic force per unit area at the base of a column of fluid. Hydrostatic gauge measurements are independent of the type of gas being measured, and can be designed to have a very linear calibration. They have poor dynamic response.

Piston

Piston-type gauges counterbalance the pressure of a fluid with a spring (for example tire-pressure gauges of comparatively low accuracy) or a solid weight, in which case it is known as a deadweight tester and may be used for calibration of other gauges.

Liquid Column (Manometer)

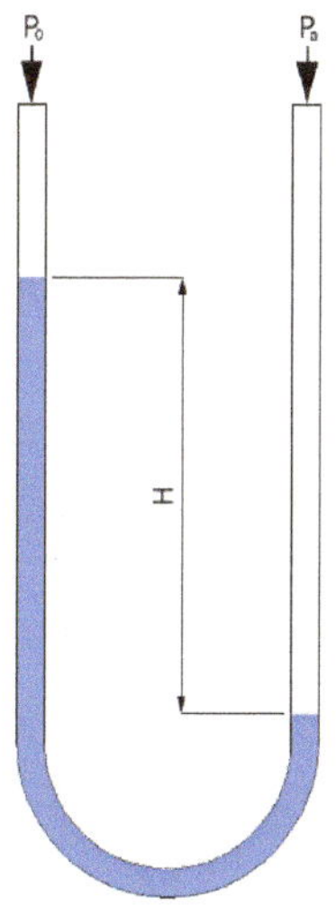

The difference in fluid height in a liquid column manometer is proportional to the pressure difference.

By using the pressure head equation, liquids can be used for instrumentation where gravity is present. Liquid column gauges consist of a vertical column of liquid in a tube that has ends which are exposed to different pressures. The column will rise or fall until its weight (a force applied due to gravity) is in equilibrium with the pressure differential between the two ends of the tube (a force applied due to fluid pressure). A very simple version is a U-shaped tube half-full of liquid, one side of which is connected to the region of interest while the reference pressure (which might be the atmospheric pressure or a vacuum) is applied to the other. The difference in liquid level represents the applied pressure. The pressure exerted by a column of fluid of height h and density ρ is given by the hydrostatic pressure equation, $P = hg\rho$. Therefore, the pressure difference between the applied pressure P_a and the reference pressure P_0 in a U-tube manometer can be found by solving $P_a - P_0 = hg\rho$. In other words, the pressure on either end of the liquid (shown in blue in the figure) must be balanced (since the liquid is static) and so $P_a = P_0 + hg\rho$.

In most liquid column measurements, the result of the measurement is the height, h, expressed typically in mm, cm, or inches. The h is also known as the pressure head. When expressed as a pressure head, pressure is specified in units of length and the measurement fluid must be specified. When accuracy is critical, the temperature of the measurement fluid must likewise be specified, because liquid density is a function of temperature. So, for example, pressure head might be written "742.2 mm_{Hg}" or "4.2 in_{H2O} at 59 °F" for measurements taken with mercury or water as the manometric fluid, respectively. The word "gauge" or "vacuum" may be added to such a measurement to distinguish between a pressure above or below the atmospheric pressure. Both mm of mercury and inches of water are common pressure heads which can be converted to S.I. units of pressure using unit conversion and the above formulas.

If the fluid being measured is significantly dense, hydrostatic corrections may have to be made for the height between the moving surface of the manometer working fluid and the location where the pressure measurement is desired except when measuring differential pressure of a fluid (for example across an orifice plate or venturi), in which case the density ρ should be corrected by subtracting the density of the fluid being measured.

Although any fluid can be used, mercury is preferred for its high density (13.534 g/cm³) and low vapour pressure. For low pressure differences, light oil or water are commonly used (the latter giving rise to units of measurement such as inches water gauge and millimetres H_2O. Liquid-column pressure gauges have a highly linear calibration. They have poor dynamic response because the fluid in the column may react slowly to a pressure change.

When measuring vacuum, the working liquid may evaporate and contaminate the vacuum if its vapor pressure is too high. When measuring liquid pressure, a loop filled with gas or a light fluid can isolate the liquids to prevent them from mixing but this can be unnecessary, for example when mercury is used as the manometer fluid to measure differential pressure of a fluid such as water. Simple hydrostatic gauges can measure pressures ranging from a few Torr (a few 100 Pa) to a few atmospheres. (Approximately 1,000,000 Pa)

A single-limb liquid-column manometer has a larger reservoir instead of one side of the U-tube and has a scale beside the narrower column. The column may be inclined to further amplify the liquid movement. Based on the use and structure following type of manometers are used

1. Simple Manometer
2. Micromanometer
3. Differential manometer
4. Inverted differential manometer

McLeod Gauge

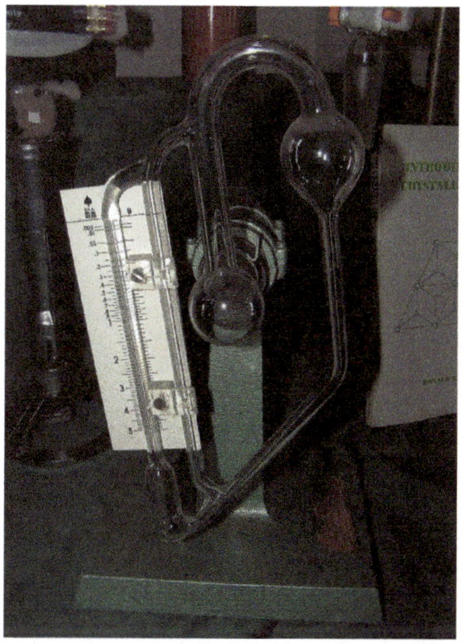

A McLeod gauge, drained of mercury

A McLeod gauge isolates a sample of gas and compresses it in a modified mercury manometer until the pressure is a few millimetres of mercury. The technique is slow and unsuited to continual monitoring, but is capable of good accuracy. Unlike other manometer gauges, the McLeod gauge reading is dependent on the composition of the gas since the interpretation relies on the sample compressing as an ideal gas. Due to the compression process, the McLeod gauge completely ignores partial pressures from non-ideal vapors that condense, such as pump oils, mercury, and even water if compressed enough.

Useful range: from around 10^{-4} torr (roughly 10^{-2} Pa) to vacuums as high as 10^{-6} Torr (0.1 mPa),

0.1 mPa is the lowest direct measurement of pressure that is possible with current technology. Other vacuum gauges can measure lower pressures, but only indirectly by measurement of other pressure-controlled properties. These indirect measurements must be calibrated to SI units via a direct measurement, most commonly a McLeod gauge.

Aneroid

Aneroid gauges are based on a metallic pressure-sensing element that flexes elastically under the effect of a pressure difference across the element. "Aneroid" means "without fluid," and the term originally distinguished these gauges from the hydrostatic gauges described above. However, aneroid gauges can be used to measure the pressure of a liquid as well as a gas, and they are not the only type of gauge that can operate without fluid. For this reason, they are often called mechanical gauges in modern language. Aneroid gauges are not dependent on the type of gas being measured, unlike thermal and ionization gauges, and are less likely to contaminate the system than hydrostatic gauges. The pressure sensing element may be a Bourdon tube, a diaphragm, a capsule, or a set of bellows, which will change shape in response to the pressure of the region in question. The deflection of the pressure sensing element may be read by a linkage connected to a needle, or it may be read by a secondary transducer. The most common secondary transducers in modern vacuum gauges measure a change in capacitance due to the mechanical deflection. Gauges that rely on a change in capacitance are often referred to as capacitance manometers.

Bourdon

Membrane-type manometer

The Bourdon pressure gauge uses the principle that a flattened tube tends to straighten or regain its circular form in cross-section when pressurized. Although this change in cross-section may be hardly noticeable, and thus involving moderate stresses within the elastic range of easily workable materials, the strain of the material of the tube is magnified by forming the tube into a C shape or even a helix, such that the entire tube tends to straighten out or uncoil, elastically, as it is pressurized. Eugene Bourdon patented his gauge in France in 1849, and it was widely adopted because of its superior sensitivity, linearity, and accuracy; Edward Ashcroft purchased Bourdon's American patent rights in 1852 and became a major manufacturer of gauges. Also in 1849, Bernard Schaeffer in Magdeburg, Germany patented a successful diaphragm pressure gauge, which, together with the Bourdon gauge, revolutionized pressure measurement in industry. But in 1875 after Bourdon's patents expired, his company Schaeffer and Budenberg also manufactured Bourdon tube gauges.

An original 19th century Eugene Bourdon compound gauge, reading pressure both below and above ambient with great sensitivity.

In practice, a flattened thin-wall, closed-end tube is connected at the hollow end to a fixed pipe containing the fluid pressure to be measured. As the pressure increases, the closed end moves in an arc, and this motion is converted into the rotation of a (segment of a) gear by a connecting link that is usually adjustable. A small-diameter pinion gear is on the pointer shaft, so the motion is magnified further by the gear ratio. The positioning of the indicator card behind the pointer, the initial pointer shaft position, the linkage length and initial position, all provide means to calibrate the pointer to indicate the desired range of pressure for variations in the behavior of the Bourdon tube itself. Differential pressure can be measured by gauges containing two different Bourdon tubes, with connecting linkages.

Bourdon tubes measure gauge pressure, relative to ambient atmospheric pressure, as opposed to absolute pressure; vacuum is sensed as a reverse motion. Some aneroid barometers use Bourdon tubes closed at both ends. When the measured pressure is rapidly pulsing, such as when the gauge is near a reciprocating pump, an orifice restriction in the connecting pipe is frequently used to avoid unnecessary wear on the gears and provide an average reading; when the whole gauge is subject to mechanical vibration, the entire case including the pointer and indicator card can be filled with an oil or glycerin. Tapping on the face of the gauge is not recommended as it will tend to falsify actual readings initially presented by the gauge. The Bourdon tube is separate from the face of the gauge and thus has no effect on the actual reading of pressure. Typical high-quality modern gauges provide an accuracy of ±2% of span, and a special high-precision gauge can be as accurate as 0.1% of full scale.

In the following illustrations the transparent cover face of the pictured combination pressure and vacuum gauge has been removed and the mechanism removed from the case. This particular gauge is a combination vacuum and pressure gauge used for automotive diagnosis:

Indicator side with card and dial

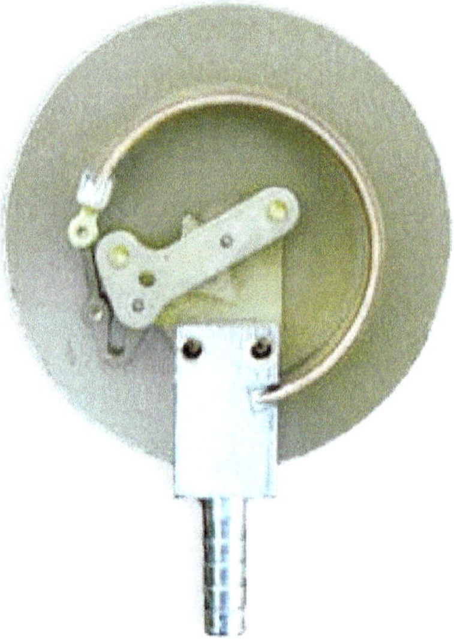

Mechanical side with Bourdon tube

- the left side of the face, used for measuring manifold vacuum, is calibrated in centimetres of mercury on its inner scale and inches of mercury on its outer scale.
- the right portion of the face is used to measure fuel pump pressure or turbo boost and is calibrated in fractions of 1 kgf/cm^2 on its inner scale and pounds per square inch on its outer scale.

Mechanical Details

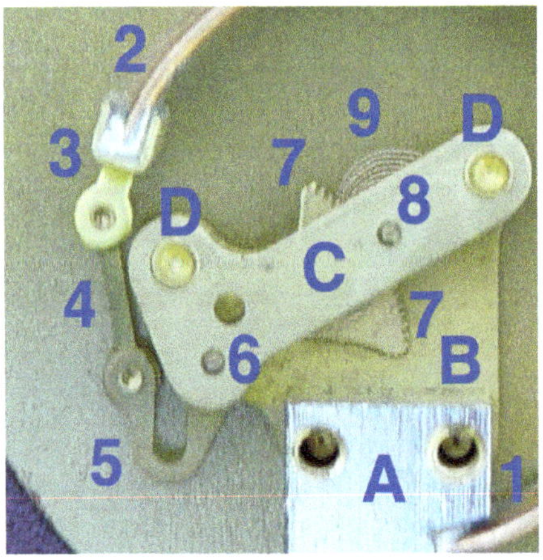

Mechanical details

Stationary parts:

- A: Receiver block. This joins the inlet pipe to the fixed end of the Bourdon tube (1) and secures the chassis plate (B). The two holes receive screws that secure the case.
- B: Chassis plate. The face card is attached to this. It contains bearing holes for the axles.
- C: Secondary chassis plate. It supports the outer ends of the axles.

- D: Posts to join and space the two chassis plates.

Moving Parts:

1. Stationary end of Bourdon tube. This communicates with the inlet pipe through the receiver block.
2. Moving end of Bourdon tube. This end is sealed.
3. Pivot and pivot pin.
4. Link joining pivot pin to lever (5) with pins to allow joint rotation.
5. Lever. This is an extension of the sector gear (7).
6. Sector gear axle pin.
7. Sector gear.
8. Indicator needle axle. This has a spur gear that engages the sector gear (7) and extends through the face to drive the indicator needle. Due to the short distance between the lever arm link boss and the pivot pin and the difference between the effective radius of the sector gear and that of the spur gear, any motion of the Bourdon tube is greatly amplified. A small motion of the tube results in a large motion of the indicator needle.
9. Hair spring to preload the gear train to eliminate gear lash and hysteresis.

Diaphragm

A second type of aneroid gauge uses deflection of a flexible membrane that separates regions of different pressure. The amount of deflection is repeatable for known pressures so the pressure can be determined by using calibration. The deformation of a thin diaphragm is dependent on the difference in pressure between its two faces. The reference face can be open to atmosphere to measure gauge pressure, open to a second port to measure differential pressure, or can be sealed against a vacuum or other fixed reference pressure to measure absolute pressure. The deformation can be measured using mechanical, optical or capacitive techniques. Ceramic and metallic diaphragms are used.

Useful range: above 10^{-2} Torr (roughly 1 Pa)

For absolute measurements, welded pressure capsules with diaphragms on either side are often used.

shape:

- Flat
- corrugated
- flattened tube
- capsule

Bellows

A pile of pressure capsules with corrugated diaphragms in an aneroid barograph

In gauges intended to sense small pressures or pressure differences, or require that an absolute pressure be measured, the gear train and needle may be driven by an enclosed and sealed bellows chamber, called an aneroid, which means "without liquid". (Early barometers used a column of liquid such as water or the liquid metal mercury suspended by a vacuum.) This bellows configuration is used in aneroid barometers (barometers with an indicating needle and dial card), altimeters, altitude recording barographs, and the altitude telemetry instruments used in weather balloon radiosondes. These devices use the sealed chamber as a reference pressure and are driven by the external pressure. Other sensitive aircraft instruments such as air speed indicators and rate of climb indicators (variometers) have connections both to the internal part of the aneroid chamber and to an external enclosing chamber.

Magnetic Coupling

These gauges use the attraction of two magnets to translate differential pressure into motion of a dial pointer. As differential pressure increases, a magnet attached to either a piston or rubber diaphragm moves. A rotary magnet that is attached to a pointer then moves in unison. To create different pressure ranges, the spring rate can be increased or decreased.

Spinning Rotor Gauge

The spinning rotor gauge works by measuring the amount a rotating ball is slowed by the viscosity of the gas being measured. The ball is made of steel and is magnetically levitated inside a steel tube closed at one end and exposed to the gas to be measured at the other. The ball is brought up to speed (about 2500 rad/s), and the speed measured after switching off the drive, by electromagnetic transducers. The range of the instrument is 10^{-5} to 10^{2} Pa (10^{3} Pa with less accuracy). It is accurate and stable enough to be used as a secondary standard. The instrument requires some skill and knowledge to use correctly. Various corrections must be applied and the ball must be spun at a pressure well below the intended measurement pressure for five hours before using. It is most useful in calibration and research laboratories where high accuracy is required and qualified technicians are available.

Electronic Pressure Sensors

Piezoresistive Strain Gage

Uses the piezoresistive effect of bonded or formed strain gauges to detect strain due to applied pressure.

Capacitive

Uses a diaphragm and pressure cavity to create a variable capacitor to detect strain due to applied pressure.

Magnetic

Measures the displacement of a diaphragm by means of changes in inductance (reluctance), LVDT, Hall Effect, or by eddy current principle.

Piezoelectric

Uses the piezoelectric effect in certain materials such as quartz to measure the strain upon the sensing mechanism due to pressure.

Optical

Uses the physical change of an optical fiber to detect strain due to applied pressure.

Potentiometric

Uses the motion of a wiper along a resistive mechanism to detect the strain caused by applied pressure.

Resonant

Uses the changes in resonant frequency in a sensing mechanism to measure stress, or changes in gas density, caused by applied pressure.

Thermal Conductivity

Generally, as a real gas increases in density -which may indicate an increase in pressure- its ability to conduct heat increases. In this type of gauge, a wire filament is heated by running current through it. A thermocouple or resistance thermometer (RTD) can then be used to measure the temperature of the filament. This temperature is dependent on the rate at which the filament loses heat to the surrounding gas, and therefore on the thermal conductivity. A common variant is the Pirani gauge, which uses a single platinum filament as both the heated element and RTD. These gauges are accurate from 10^{-3} Torr to 10 Torr, but their calibration is sensitive to the chemical composition of the gases being measured.

Pirani (One Wire)

A Pirani gauge consist of a metal wire open to the pressure being measured. The wire is heated by a current flowing through it and cooled by the gas surrounding it. If the gas pressure is reduced, the cooling effect will decrease, hence the equilibrium temperature of the wire will increase. The resistance of the wire is a function of its temperature: by measuring the voltage across the wire and the current flowing through it, the resistance (and so the gas pressure) can be determined. This type of gauge was invented by Marcello Pirani.

Two-wire

In two-wire gauges, one wire coil is used as a heater, and the other is used to measure temperature due to convection. Thermocouple gauges and thermistor gauges work in this manner using thermocouple or thermistor, respectively, to measure the temperature of the heated wire.

Ionization Gauge

Ionization gauges are the most sensitive gauges for very low pressures (also referred to as hard or high vacuum). They sense pressure indirectly by measuring the electrical ions produced when the gas is bombarded with electrons. Fewer ions will be produced by lower density gases. The calibration of an ion gauge is unstable and dependent on the nature of the gases being measured, which is not always known. They can be calibrated against a McLeod gauge which is much more stable and independent of gas chemistry.

Thermionic emission generate electrons, which collide with gas atoms and generate positive ions. The ions are attracted to a suitably biased electrode known as the collector. The current in the collector is proportional to the rate of ionization, which is a function of the pressure in the system. Hence, measuring the collector current gives the gas pressure. There are several sub-types of ionization gauge.

Useful range: 10^{-10} - 10^{-3} torr (roughly 10^{-8} - 10^{-1} Pa)

Most ion gauges come in two types: hot cathode and cold cathode. In the hot cathode version, an electrically heated filament produces an electron beam. The electrons travel through the gauge and ionize gas molecules around them. The resulting ions are collected at a negative electrode. The current depends on the number of ions, which depends on the pressure in the gauge. Hot cathode gauges are accurate from 10^{-3} Torr to 10^{-10} Torr. The principle behind cold cathode version is the same, except that electrons are produced in the discharge of a high voltage. Cold Cathode gauges are accurate from 10^{-2} Torr to 10^{-9} Torr. Ionization gauge calibration is very sensitive to construction geometry, chemical composition of gases being measured, corrosion and surface deposits. Their calibration can be invalidated by activation at atmospheric pressure or low vacuum. The composition of gases at high vacuums will usually be unpredictable, so a mass spectrometer must be used in conjunction with the ionization gauge for accurate measurement.

Hot Cathode

A hot-cathode ionization gauge is composed mainly of three electrodes acting together as a triode, wherein the cathode is the filament. The three electrodes are a collector or plate, a filament, and a grid. The collector current is measured in picoamps by an electrometer. The filament voltage to ground is usually at a potential of 30 volts, while the grid voltage at 180–210 volts DC, unless there is an optional electron bombardment feature, by heating the grid, which may have a high potential of approximately 565 volts. The most common ion gauge is the hot-cathode Bayard-Alpert gauge, with a small ion collector inside the grid. A glass envelope with an opening to the vacuum can surround the electrodes, but usually the Nude Gauge is inserted in the vacuum chamber directly, the pins being fed through a ceramic plate in the wall of the chamber. Hot-cathode gauges can be damaged or lose their calibration if they are exposed to atmospheric pressure or even low vacuum while hot. The measurements of a hot-cathode ionization gauge are always logarithmic.

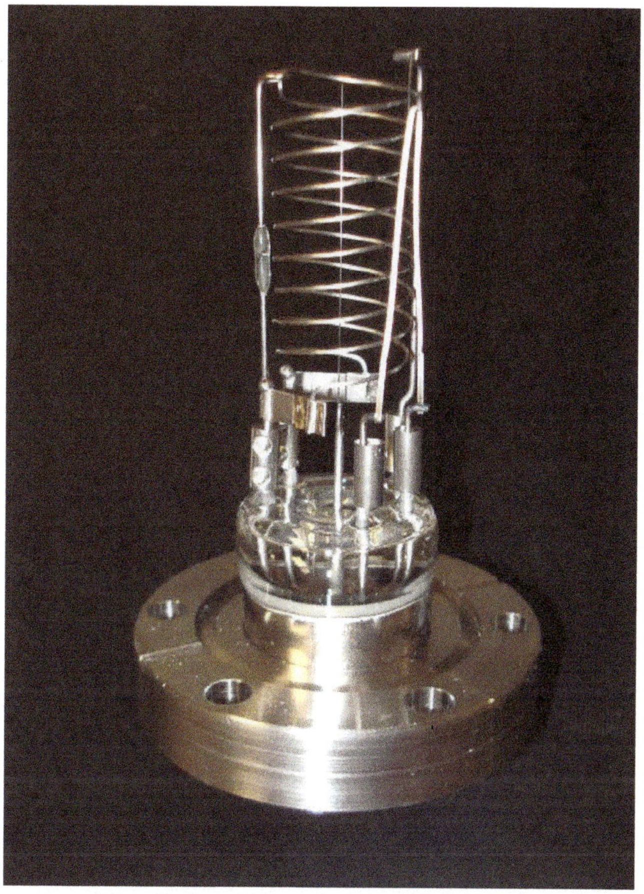

Bayard-Alpert hot-cathode ionization gauge

Electrons emitted from the filament move several times in back and forth movements around the grid before finally entering the grid. During these movements, some electrons collide with a gaseous molecule to form a pair of an ion and an electron (Electron ionization). The number of these ions is proportional to the gaseous molecule density multiplied by the electron current emitted from the filament, and these ions pour into the collector to form an ion current. Since the gaseous molecule density is proportional to the pressure, the pressure is estimated by measuring the ion current.

The low-pressure sensitivity of hot-cathode gauges is limited by the photoelectric effect. Electrons hitting the grid produce x-rays that produce photoelectric noise in the ion collector. This limits the range of older hot-cathode gauges to 10^{-8} Torr and the Bayard-Alpert to about 10^{-10} Torr. Additional wires at cathode potential in the line of sight between the ion collector and the grid prevent this effect. In the extraction type the ions are not attracted by a wire, but by an open cone. As the ions cannot decide which part of the cone to hit, they pass through the hole and form an ion beam. This ion beam can be passed on to a:

- Faraday cup
- Microchannel plate detector with Faraday cup
- Quadrupole mass analyzer with Faraday cup
- Quadrupole mass analyzer with Microchannel plate detector Faraday cup
- ion lens and acceleration voltage and directed at a target to form a sputter gun. In this case a valve lets gas into the grid-cage.

Cold Cathode

There are two subtypes of cold-cathode ionization gauges: the Penning gauge (invented by Frans Michel Penning), and the Inverted magnetron, also called a Redhead gauge. The major difference between the two is the position of the anode with respect to the cathode. Neither has a filament, and each may require a DC potential of about 4 kV for operation. Inverted magnetrons can measure down to $1x10^{-12}$ Torr.

Likewise, cold-cathode gauges may be reluctant to start at very low pressures, in that the near-absence of a gas makes it difficult to establish an electrode current - in particular in Penning gauges, which use an axially symmetric magnetic field to create path lengths for electrons that are of the order of metres. In ambient air, suitable ion-pairs are ubiquitously formed by cosmic radiation; in a Penning gauge, design features are used to ease the set-up of a discharge path. For example, the electrode of a Penning gauge is usually finely tapered to facilitate the field emission of electrons.

Maintenance cycles of cold cathode gauges are, in general, measured in years, depending on the gas type and pressure that they are operated in. Using a cold cathode gauge in gases with substantial organic components, such as pump oil fractions, can result in the growth of delicate carbon films and shards within the gauge that eventually either short-circuit the electrodes of the gauge or impede the generation of a discharge path.

Dynamic Transients

When fluid flows are not in equilibrium, local pressures may be higher or lower than the average pressure in a medium. These disturbances propagate from their source as longitudinal pressure variations along the path of propagation. This is also called sound. Sound pressure is the instantaneous local pressure deviation from the average pressure caused by a sound wave. Sound pressure can be measured using a microphone in air and a hydrophone in water. The effective sound pressure is the root mean square of the instantaneous sound pressure over a given interval of time. Sound pressures are normally small and are often expressed in units of microbar.

- frequency response of pressure sensors
- resonance

Calibration and Standards

Dead-weight tester. This uses known calibrated weights on a piston to generate a known pressure.

The American Society of Mechanical Engineers (ASME) has developed two separate and distinct standards on pressure Measurement, B40.100 and PTC 19.2. B40.100 provides guidelines on Pressure Indicated Dial Type and Pressure Digital Indicating Gauges, Diaphragm Seals, Snubbers, and Pressure Limiter Valves. PTC 19.2 provides instructions and guidance for the accurate determination of pressure values in support of the ASME Performance Test Codes. The choice of method, instruments, required calculations, and corrections to be applied depends on the purpose of the measurement, the allowable uncertainty, and the characteristics of the equipment being tested.

The methods for pressure measurement and the protocols used for data transmission are also provided. Guidance is given for setting up the instrumentation and determining the uncertainty of the measurement. Information regarding the instrument type, design, applicable pressure range, accuracy, output, and relative cost is provided. Information is also provided on pressure-measuring devices that are used in field environments i.e., Piston Gauges, Manometers, and Low-Absolute-Pressure (Vacuum) Instruments.

These methods are designed to assist in the evaluation of measurement uncertainty based on current technology and engineering knowledge, taking into account published instrumentation specifications and measurement and application techniques. This Supplement provides guidance in the use of methods to establish the pressure-measurement uncertainty.

European (CEN) Standard

- EN 472 : Pressure gauge - Vocabulary.
- EN 837-1 : Pressure gauges. Bourdon tube pressure gauges. Dimensions, metrology, requirements and testing.
- EN 837-2 : Pressure gauges. Selection and installation recommendations for pressure gauges.
- EN 837-3 : Pressure gauges. Diaphragm and capsule pressure gauges. Dimensions, metrology, requirements, and testing.

US ASME Standards

- B40.100-2013: Pressure gauges and Gauge attachments.
- PTC 19.2-2010 : Performance test code for pressure measurement.

Pressure measurement is important in many fluid mechanics related applications. From appropriate pressure measurements velocity, aerodynamic forces and moments can be determined. Pressure is measured by the force acting on unit area. Measuring devices usually indicate differential pressure i.e. in relation with atmospheric pressure. This is called gauge pressure. The measured pressure may be positive or negative with reference to the atmospheric pressure .A negative gauge pressure is referred to as vacuum.

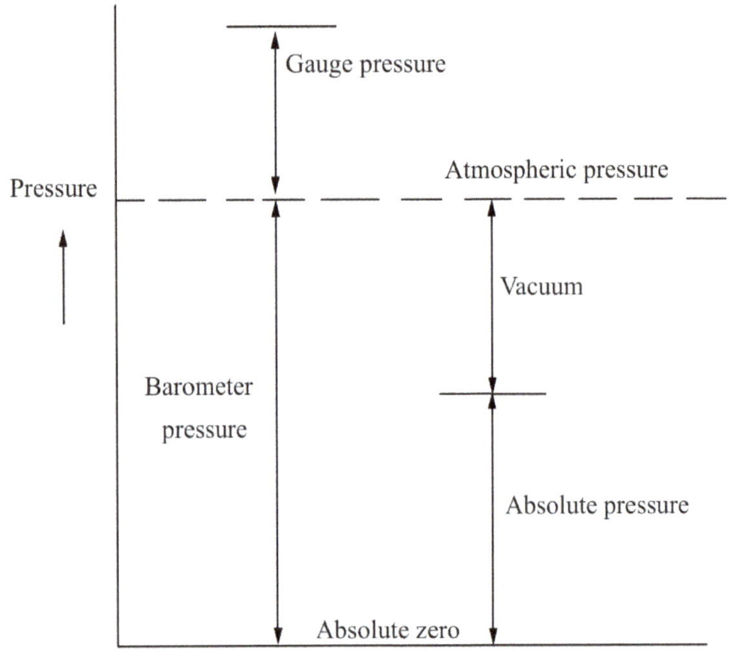

Explanation of the pressure terminology

Units of Pressure

1Pascal(1N/m^2)	= 10dyne/cm^2
1mmHg	= 133.32pascals
	= 13.595mm Water
Standard atmosphere	= 1.013 * 10^5N/m^2
1 millibar	= 1000 dyn/cm^2
1 micron	= 10^{-6}mHg
1 torr	= 1 mmHg
	= 1000micron

Absolute Pressure

Absolute pressure is determined as algebraic sum of the readings of a barometer and of a manometer showing the gauge pressure. Manometers which measure absolute pressure are also available. They measure the pressure with reference to absolute zero pressure.

Pressure Measuring Devices

Main characteristics of manometers are pressure range, accuracy, sensitivity and speed of response. Pressure range of manometers varies from almost perfect vacuum to several hundreds of atmosphere. The conventional instruments used for pressure measurement are divided into the following groups.

1. Liquid column manometers
2. Pressure gauges with elastic sensing elements
3. Pressure transducers
4. Manometers for low absolute pressures
5. Manometers for very high absolute pressures

Liquid Column Manometers

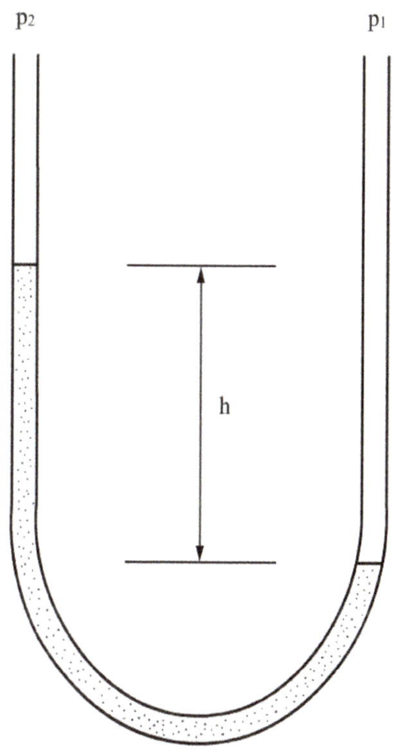

Liquid column manometer

For amplifying the deflection in a liquid column manometer, liquids with lower density could be used or one of the limbs of the manometer may be inclined. Commonly used manometric liquids are mercury, water or alcohol. Some of the important and desirable properties of the manometric liquids are:

- High chemical stability
- Low viscosity
- Low capillary constant
- Low coefficient of thermal expansion
- Low volatility
- Low vapour pressure

High thermal stability and low volatility are important for maintaining a constant. Specific gravity. High viscosity causes transmission lags .Thermal expansion causes changes in zero reading. While measuring low pressures, vapour pressure of the manometric fluid is an important consideration. Properties of some of the commonly used fluids are given in the table below.

Table: Typical properties of manometric fluids

Fluid	Specific gravity	B.P.(^{o}C) at 760mm Hg	Surface Tension dyn/cm	Viscosity Cp	Coeft.of Volumetric Expansion *10^5
Methyl Alcohol	0.792	64.7	22.6	0.59	
Ethyl Alcohol	0.789	78.4	22.0	1.9	110
Mercury	13.55	356.59	465	1.55	18
Toluene	0.866	110.8	28.4		
CCl_4	1.594	76.8	26.8	0.97	

Inclined Manometer

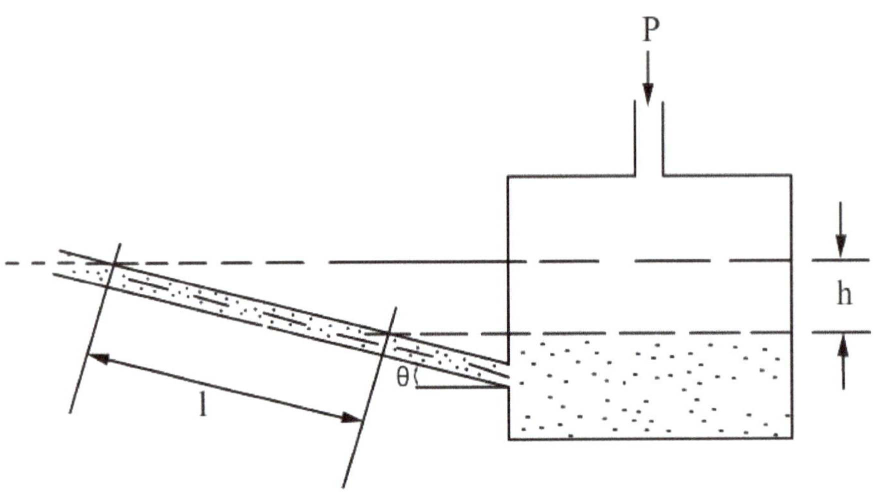

Inclined manometer

The figure above shows the amplification of the reading of applied pressure 'h' as

$$\frac{h}{l} = \sin\theta$$

Mercury Barometer

Barometer is the device used to measure the atmospheric pressure. Mercury barometer consists essentially of a glass tube sealed at one end and mounted vertically in a bowl or cistern of mercury so that the open end of the tube is submerged below the surface of mercury in the cistern. In the figure below (i) the zero level of mercury in the cistern is shown under the influence of atmospheric pressure p_1. When $p_2=p_1$ = atmospheric pressure, the zero level in the cistern can be marked. If the tube is open and if different pressures act in the cistern and in the tube, then there will be difference in the levels of mercury. If p_1 is greater than p_2 as shown in the figure below (ii), then mercury will be forced down in the cistern and corresponding rise will be there in the tube.

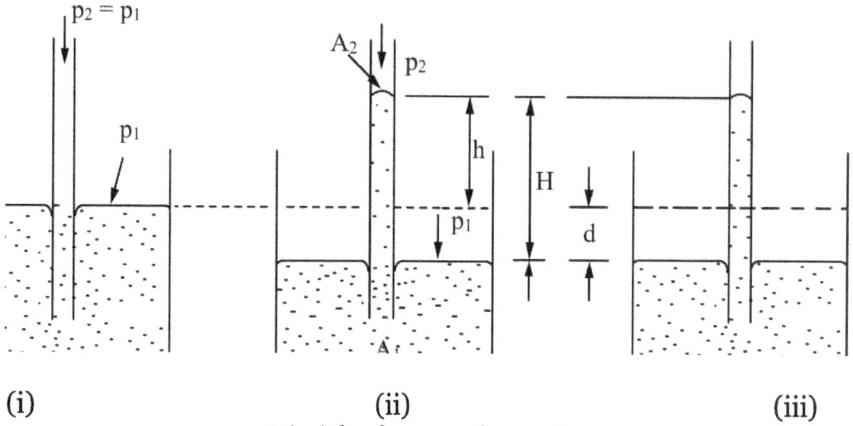

Principle of mercury barometer

The balance of pressures will be given by $p_1 = p_2 + H\rho$ 1

(where H is the difference between levels of mercury in the cistern and in the tube and ρ the density of mercury). From the figure above (ii), H=h+d

The quantity of mercury that has left the cistern is same as that has risen in the tube.

$A_1 d = A_2 h$ where A_1 and A_2 are areas of the cistern and the tube respectively.

$$d = \frac{A_2}{A_1} h$$

$$\therefore H = h + h\left(\frac{A_2}{A_1}\right)$$

$$= h\left(1 + \frac{A_2}{A_1}\right) \quad \text{............................... 2}$$

Replacing H in equation 1

$$p_1 = p_2 + h\left(1 + \frac{A_2}{A_1}\right)\rho \quad \text{.......................3}$$

$p_2 \to 0$ in practical cases as the tube will be evacuated and sealed.

$$p_1 = h\left(1 + \frac{A_2}{A_1}\right)\rho \quad \text{.......................................4}$$

Micromanometer

For accurate measurement of extremely small pressure differences micromanometers are used. In the figure given below, the instrument is initially adjusted such that $p_1 = p_2$.

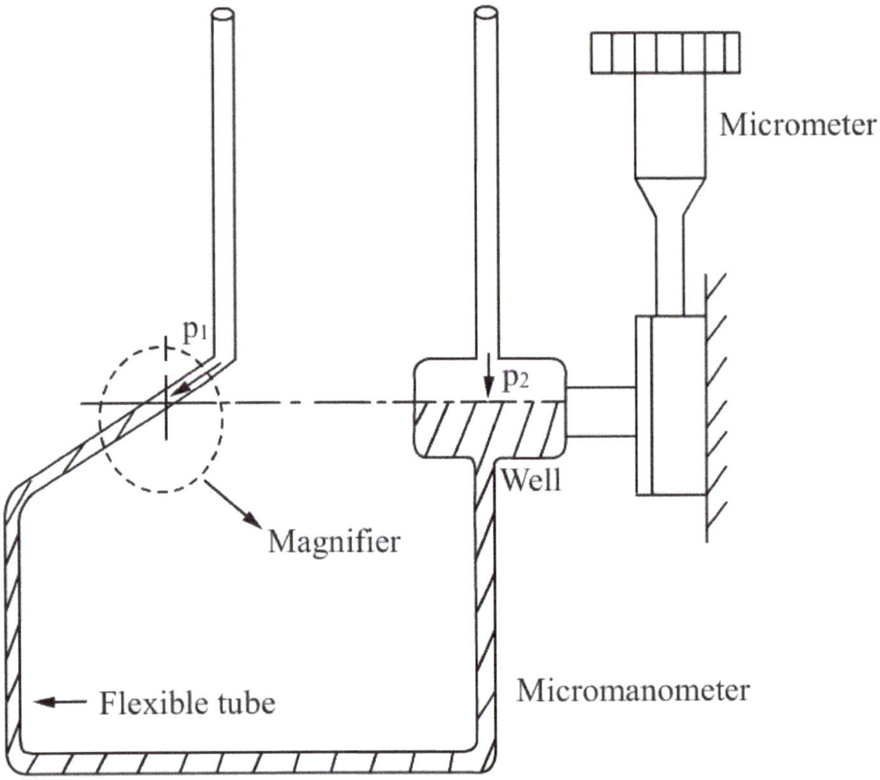

Typical Micromanometer

The meniscus in the inclined tube is located at a reference level fixed by the hairline viewed through the magnifier. The reading of the micrometer is noted. Application of the unknown pressure difference causes the meniscus to move off the hairline but it can be restored to the initial position by raising or lowering the well (mercury sump) The difference in the initial and final micrometer readings gives the height of the mercury column and hence the pressure. Pressures as low as 0.025mm water column can be measured.

Mechanical Manometers

Mechanical manometers provide faster response than liquid column manometers. In liquid column measurements, lag is due to the displacements of the liquid. In elastic sensing element type of manometers the time lag is due to the time required for equalisation of pressure to be measured with that in the sensing chamber. The deformation of elastic sensing elements is measured with the aid of kinematic, optical or electrical systems. There are three types of elastic sensing elements which are (i) Bourdon Tubes (ii) Diaphragms (flat or corrugated) (iii) Bellows

Bourdon Tube

Bourdon tube is the oldest pressure sensing element. It is a length of metal tube of elliptical cross section and shaped into letter 'C'.

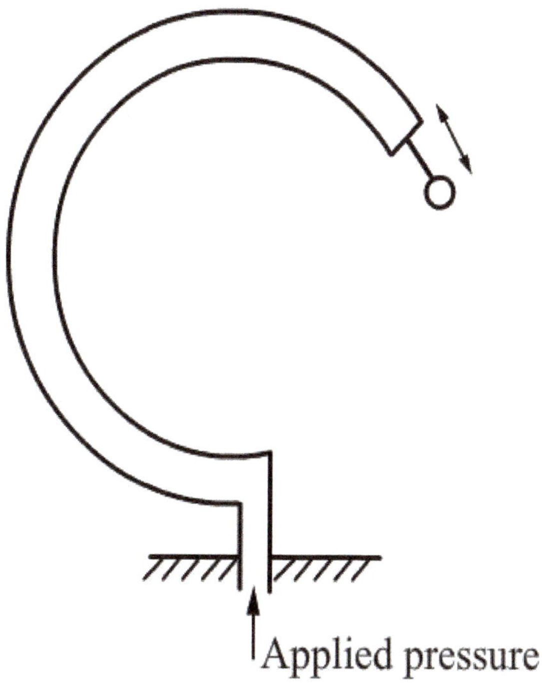

Bourdon tube pressure gauge

A tube of elliptical cross section has a smaller volume than a circular one of the same length and perimeter. When connected to the pressure source it is made to accommodate more of the fluid. Resultant of all reactions will produce maximum displacement at the free end. Within close limits, the change in angle subtended at the centre by a tube is proportional to the change of internal pressure and within the limits of proportionality of the material; the displacement of the free end is proportional to the applied pressure.

The ratio between major and minor axes decides the sensitivity of the Bourdon tube. The larger is the higher is the sensitivity. Materials of the Bourdon tube is Phosphor bronze, Beryllium bronze or Beryllium Copper.

Elastic Diaphragms

The pressure sensors making use of elastic diaphragms consist of a diaphragm fixed in a tubular member. The pressure to be measured is applied on one side. The mathematical relation between pressure and central deflection for a flat circular diaphragm is given by

$$p = \frac{16Et^4}{3a^4\left(1-\mu^2\right)}\left[\frac{y_c}{t} + 0.488\left(\frac{y_c}{t}\right)^3 + ...\right] \text{..........................5}$$

To have a linear pressure deflection relation, the second and later terms should be small.

p = applied pressure

t = thickness

a = radius

yc = central deflection

E = Youngs modulus

μ = Poisson's ratio $\rightarrow \frac{\textit{lateral strain}}{\textit{Axial strain}}$

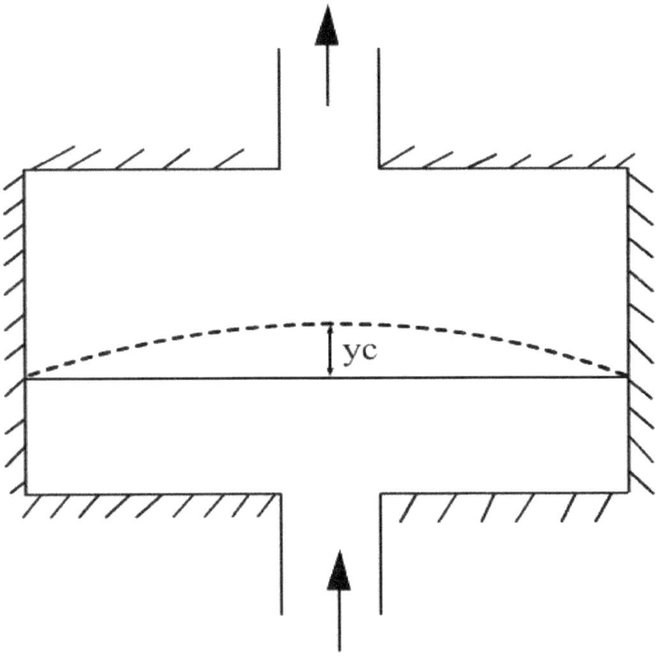

Diaphragm pressure gauge

As seen from the equation the pressure deflection relation is not linear. If a non-linearity of 5% is acceptable the deflection must be less than 1/3 of thickness. Neglecting the higher order terms within brackets,

$$p = \frac{16Et^4}{3a^4\Pi(1-\mu^2)} \times \frac{y_c}{t}$$

$$p = \frac{16Et^3 y_c}{3a^4(1-\mu^2)}$$

$$\frac{y_c}{p} = \frac{y_c 3a^4(1-\mu^2)}{16Et^3 y_c} \quad \ldots\ldots 6$$

$$\frac{y_c}{p} = K = \frac{3(1-\mu^2)a^4}{16Et^3} \quad \ldots\ldots 7$$

where 'K' is the sensitivity of the device.

The natural frequency of the diaphragm may be written as follows:

$$\omega = 10.2\left[\frac{E}{12p_d\left(1-\mu^2\right)}\right]^{1/2}\frac{t}{a^2}\, rad/s \quad ....8$$

Diaphragm has infinite number of natural frequencies and the lowest of them is considered here. From the two equations for sensitivity 'K' and the natural frequency ω.

Sensitivity (K) and natural frequency (ω) are related by

$$\frac{y_c}{p} = K = \frac{1.63}{\omega^2 t \rho_d} \quad ....9$$

(ρ_d is density of diaphragm material)

Sensitivity is inversely proportional to square of natural frequency. Sensitivity can be improved by lowering the natural frequency. The natural frequency should be 3 to 4 times the frequency of pressure pulsations. Compared to bourdon tubes diaphragms have higher natural frequencies. Hence, they can be used for the measurement of dynamic pressures of higher frequency fluctuations. They can be installed flush with the surface of the body and hence there is only much less transmission lag.

Corrugated Diaphragms

Corrugated diaphragms permit considerably larger deflections than flat diaphragms. Their number and depth control the response and sensitivity. The greater the number and depth, the more linear is its deflection and greater is the sensitivity.

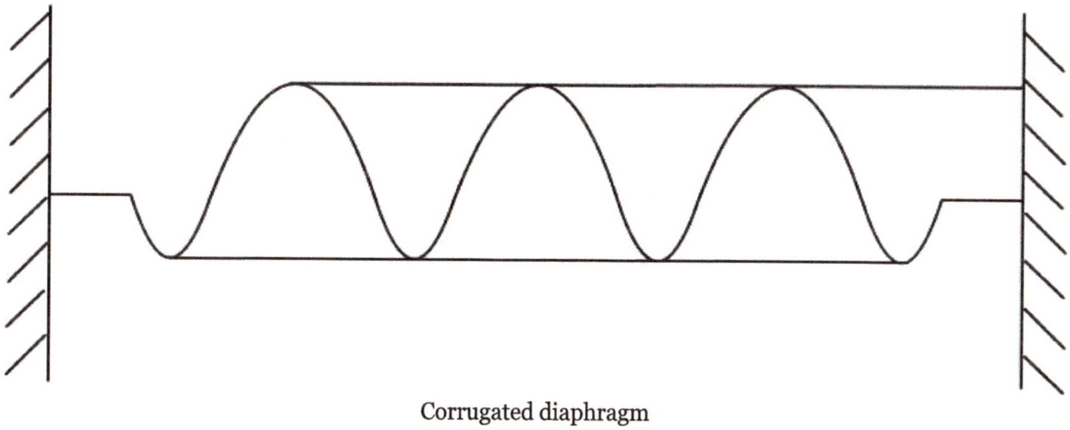

Corrugated diaphragm

Capsules, Bellows

For even larger deflections than the diaphragms corrugated diaphragms are made in to boxes or bellows. Bellows are most commonly used to measure small steady pressures.

Pressure Sensor

Digital air pressure sensor

Miniature digital barometric pressure sensor

Some pressure sensors, such as those found in some traffic enforcement cameras, function in a binary (off/on) manner, i.e., when pressure is applied to a pressure sensor, the sensor acts to complete or break an electrical circuit. These types of sensors are also known as a pressure switch.

Types of Pressure Measurements

silicon piezoresistive pressure sensors

Pressure sensors can be classified in terms of pressure ranges they measure, temperature ranges of operation, and most importantly the type of pressure they measure. Pressure sensors are variously named according to their purpose, but the same technology may be used under different names.

- Absolute pressure sensor

This sensor measures the pressure relative to perfect vacuum.

- Gauge pressure sensor

This sensor measures the pressure relative to atmospheric pressure. A tire pressure gauge is an example of gauge pressure measurement; when it indicates zero, then the pressure it is measuring is the same as the ambient pressure.

- Vacuum pressure sensor

This term can cause confusion. It may be used to describe a sensor that measures pressures below atmospheric pressure, showing the difference between that low pressure and atmospheric pressure (i.e. negative gauge pressure), but it may also be used to describe a sensor that measures low pressure relative to perfect vacuum (i.e. absolute pressure).

- Differential pressure sensor

This sensor measures the difference between two pressures, one connected to each side of the sensor. Differential pressure sensors are used to measure many properties, such as pressure drops across oil filters or air filters, fluid levels (by comparing the pressure above and below the liquid) or flow rates (by measuring the change in pressure across a restriction). Technically speaking, most pressure sensors are really differential pressure sensors; for example a gauge pressure sensor is merely a differential pressure sensor in which one side is open to the ambient atmosphere.

- Sealed pressure sensor

This sensor is similar to a gauge pressure sensor except that it measures pressure relative to some fixed pressure rather than the ambient atmospheric pressure (which varies according to the location and the weather).

Pressure-sensing Technology

There are two basic categories of analog pressure sensors,

Force collector types These types of electronic pressure sensors generally use a force collector (such a diaphragm, piston, bourdon tube, or bellows) to measure strain (or deflection) due to applied force over an area (pressure).

- *Piezoresistive strain gauge*

  Uses the piezoresistive effect of bonded or formed strain gauges to detect strain due to applied pressure, resistance increasing as pressure deforms the material. Common technology types are Silicon (Monocrystalline), Polysilicon Thin Film, Bonded Metal Foil, Thick Film, and Sputtered Thin Film. Generally, the strain gauges are connected to form a Wheatstone bridge circuit to maximize the output of the sensor and to reduce sensitivity to errors. This is the most commonly employed sensing technology for general purpose pressure measurement.

- *Capacitive*

  Uses a diaphragm and pressure cavity to create a variable capacitor to detect strain due to applied pressure, capacitance decreasing as pressure deforms the diaphragm. Common technologies use metal, ceramic, and silicon diaphragms.

- *Electromagnetic*

  Measures the displacement of a diaphragm by means of changes in inductance (reluctance), LVDT, Hall Effect, or by eddy current principle.

- *Piezoelectric*

  Uses the piezoelectric effect in certain materials such as quartz to measure the strain upon the sensing mechanism due to pressure. This technology is commonly employed for the measurement of highly dynamic pressures.

- *Optical*

  Techniques include the use of the physical change of an optical fiber to detect strain due to applied pressure. A common example of this type utilizes Fiber Bragg Gratings. This technology is employed in challenging applications where the measurement may be highly remote, under high temperature, or may benefit from technologies inherently immune to electromagnetic interference. Another analogous technique utilizes an elastic film constructed in layers that can change reflected wavelengths according to the applied pressure (strain).

- *Potentiometric*

  Uses the motion of a wiper along a resistive mechanism to detect the strain caused by applied pressure.

Other Types

These types of electronic pressure sensors use other properties (such as density) to infer pressure of a gas, or liquid.

- *Resonant*

  Uses the changes in resonant frequency in a sensing mechanism to measure stress, or changes in gas density, caused by applied pressure. This technology may be used in conjunction with a force collector, such as those in the category above. Alternatively, resonant technology may be employed by exposing the resonating element itself to the media, whereby the resonant frequency is dependent upon the density of the media. Sensors have been made out of vibrating wire, vibrating cylinders, quartz, and silicon MEMS. Generally, this technology is considered to provide very stable readings over time.

- *Thermal*

  Uses the changes in thermal conductivity of a gas due to density changes to measure pressure. A common example of this type is the Pirani gauge.

- *Ionization*

  Measures the flow of charged gas particles (ions) which varies due to density changes to measure pressure. Common examples are the Hot and Cold Cathode gauges.

Applications

There are many applications for pressure sensors:

- Pressure sensing

This is where the measurement of interest is pressure, expressed as a force per unit area. This is useful in weather instrumentation, aircraft, automobiles, and any other machinery that has pressure functionality implemented.

- Altitude sensing

This is useful in aircraft, rockets, satellites, weather balloons, and many other applications. All these applications make use of the relationship between changes in pressure relative to the altitude. This relationship is governed by the following equation:

$$h = (1 - (P / P_{\text{ref}})^{0.190284}) \times 145366.45\text{ft}$$

This equation is calibrated for an altimeter, up to 36,090 feet (11,000 m). Outside that range, an error will be introduced which can be calculated differently for each different pressure sensor. These error calculations will factor in the error introduced by the change in temperature as we go up.

Barometric pressure sensors can have an altitude resolution of less than 1 meter, which is significantly better than GPS systems (about 20 meters altitude resolution). In navigation applications altimeters are used to distinguish between stacked road levels for car navigation and floor levels in buildings for pedestrian navigation.

- Flow sensing

This is the use of pressure sensors in conjunction with the venturi effect to measure flow. Differential pressure is measured between two segments of a venturi tube that have a different aperture. The pressure difference between the two segments is directly proportional to the flow rate through the venturi tube. A low pressure sensor is almost always required as the pressure difference is relatively small.

- Level / depth sensing

A pressure sensor may also be used to calculate the level of a fluid. This technique is commonly employed to measure the depth of a submerged body (such as a diver or submarine), or level of contents in a tank (such as in a water tower). For most practical purposes, fluid level is directly proportional to pressure. In the case of fresh water where the contents are under atmospheric pressure, 1psi = 27.7 inH2o / 1Pa = 9.81 mmH2o. The basic equation for such a measurement is

$$P = \rho g h$$

where P = pressure, ρ = density of the fluid, g = standard gravity, h = height of fluid column above pressure sensor

- Leak testing

A pressure sensor may be used to sense the decay of pressure due to a system leak. This is commonly done by either comparison to a known leak using differential pressure, or by means of utilizing the pressure sensor to measure pressure change over time.

Ratiometric Correction of Transducer Output

Piezoresistive transducers configured as Wheatstone bridges often exhibit ratiometric behavior with respect not only to the measured pressure, but also the transducer supply voltage.

$$V_{\text{out}} = \frac{P \times K \times Vs_{\text{actual}}}{Vs_{\text{ideal}}}$$

where:

V_{out} is the output voltage of the transducer.

P is the actual measured pressure.

K is the nominal transducer scale factor (given an ideal transducer supply voltage) in units of voltage per pressure.

Vs_{actual} is the actual transducer supply voltage.

Vs_{ideal} is the ideal transducer supply voltage.

Correcting measurements from transducers exhibiting this behavior requires measuring the actual transducer supply voltage as well as the output voltage and applying the inverse transform of this behavior to the output signal:

$$P = \frac{V_{\text{out}} \times Vs_{\text{ideal}}}{K \times Vs_{\text{actual}}}$$

Note: Common mode signals often present in transducers configured as Wheatstone bridges are not considered in this analysis.

Pressure Transducers

Diaphragm Type Pressure Transducers

They convert the pressure to be measured into electrical signals. For example, pressure transducers whose operating principle is based on measuring changes in inductive, capacitive or ohmic resistances caused by the deformation of an elastic element.

Capacitance $C = \in_a a / d$................................10

Where $\in_a \rightarrow$ absolute permittivity

$$\frac{\in_a}{\in_0} = K \frac{1}{4\Pi \times 9 \times 10^9 E/m}$$

a = area plates

d= distance between the plates

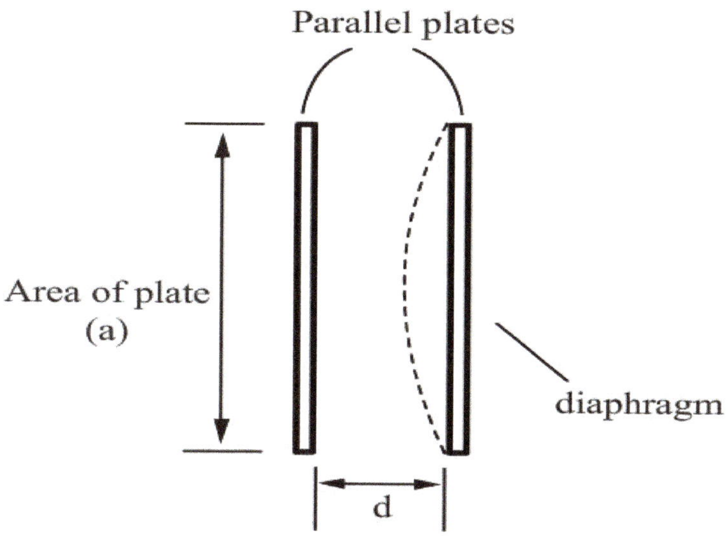

Elastic diaphragm used in a parallel plate capacitor

With the application of pressure, the diaphragm deforms with the central deflection as shown. This changes the capacitance in an electrical circuit which can be calibrated against pressure to be measured.

Piezo-Electric Pressure Transducers

The word piezo-electric is derived from the Greek word "piezein" meaning squeeze or press. Certain materials possess the ability to generate electrical potential when subjected to mechanical strain. Conversely, they change dimensions when supplied with voltage. The potential developed by application of stress is not held under static conditions. Dynamic pressures in the range of frequencies from kHz to 100MHz can be measured using piezo- electric transducers. The use of piezo-electric effect is limited to dynamic measurements. Some materials exhibiting piezo electric properties are quartz, tourmaline, barium titanate, and Lead zirconate. Quartz is the preferred material, as it possesses good mechanical properties. Also, it is a good insulator least influenced by moisture.

Pressure Sensitive Paints (PSP)

They are composed of luminescent molecules dispersed in Oxygen permeable polymer binder. When PSP is exposed to blue or an ultraviolet light the luminescent molecules are excited to a higher energy level.

From this excited state they can discharge in three ways:

a) by discharging light

b) by transferring energy to the polymer binder(heating)

c) colliding with Oxygen molecules.

Since the luminescent molecules react with Oxygen they collide and release light at the same time. The amount of light emitted is inversely proportional to the amount of Oxygen molecules on the surface.

Measurement of High Pressures

Electrical Resistance Gauges – Principle of Operation

Bourdon Tube or strain gauge can be used for high pressures. Very high pressures (say above 1000bars) may be measured by means of electrical resistance gauges which are known as Bridgeman gauge. By way of principle of operation, they make use of resistance change brought about by direct application of pressure to the conductor itself. Referring to the Figure, the sensing element is the thin wire of Magnanin (84 Cu + 12 Mn + 4 Ni) or an alloy of Gold and Chromium (2.1%) which is loosely wound. When pressure is applied, bulk compression effects produce a change in resistance which may be calibrated against pressure. The general relation between electrical and mechanical may be derived as follows:

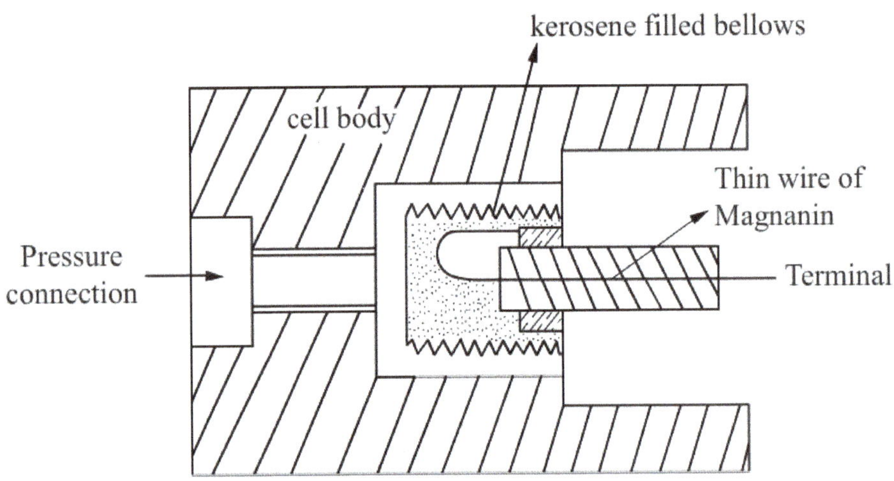

Sensor of a Bridgemans Gauge

$$R = \frac{\rho L}{A} = \frac{\rho L}{CD^2} \qquad \text{......................................11}$$

R = Resistance, ohms

L = Length of conductors

A = Area = CD^2

where D is the diameter of the circular conductor and C a constant

ρ = Resistivity ohm cm

If the conductor is strained each of the quantities in the equation will change.

Differentiating

$$dR = \frac{CD^2(Ld\rho + \rho dL) - 2C\rho DLdD}{(CD^2)^2}$$

$$= \frac{1}{CD^2}\left[(Ld\rho + \rho dL) - 2\rho L\frac{dD}{D}\right] \text{.....................12}$$

Dividing (12) by (11) gives

$$\frac{dR}{R} = \frac{dL}{L} - \frac{2dD}{D} + \frac{d\rho}{\rho}$$

The wire will be subjected to biaxial stresses because the ends in providing electrical continuity will not be subjected to pressure.

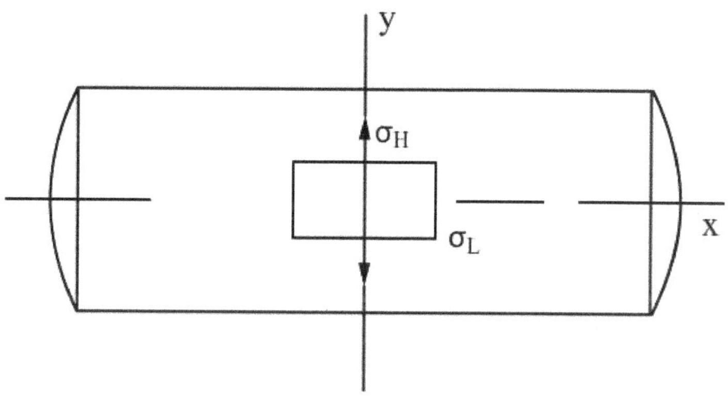

Bi-axial stresses on a small element

Let us consider a general element subjected to stresses which may be denoted as σ_x *and* σ_y. Suppose σ_x *and* σ_y are applied one at a time. If σ_x is applied first, there will be strain in the

$$\text{x- direction} = \frac{\sigma_x}{E}$$

Because of Poisson's effect, the strain in y direction $= -\mu\frac{\sigma}{E}$

If σ_y is applied first

Strain in y direction $= \frac{\sigma_y}{E}$

Strain in x direction $= -\mu\frac{\sigma_y}{E}$

There will be a net strain of

$$\in_x = \frac{1}{E}(\sigma_x - \mu\sigma_y)$$

$$\in_y = -\frac{1}{E}(\sigma y - \mu\sigma_x)$$

Stresses are nothing but applied pressure

$\sigma_x = \sigma_y = -p$ assuming $\sigma_Z = 0$

Lateral Strain

$$\in_x = \in_y = \frac{dD}{D} = -\frac{p}{E}(1-\mu)$$

$$\in_Z = \frac{dL}{L} = 2\mu\frac{p}{E}$$

Biaxial stress situation is assumed.

Longitudinal Strain

$$\therefore \frac{dR}{R} = \frac{dL}{L} - \frac{2dD}{D} + \frac{d\rho}{\rho}$$ (Substitute from above)

$$= 2\mu\frac{p}{E} + 2\frac{p}{E}(1-\mu) + \frac{d\rho}{\rho}$$

$$= \frac{2p}{E} + \frac{d\rho}{\rho}$$

If specific resistance ρ does not depend on pressure $\frac{d\rho}{\rho}$ can be neglected.

$$\frac{dR}{R} = 2\frac{p}{E}$$

R = Ro (1 + bp) where, b = 2 / E.

$$\frac{dR/R}{p} = \frac{2}{E} + \frac{d\rho/\rho}{\rho} \quad ....13$$

where b is called pressure coefficient of resistance. Resistance varies linearly with pressure.

Ranges of Different Manometers

Liquid column manometers	10 – 0.5 x 10^6 pascals
Bourdon tubes	Vacuum to a few thousand bars
Diaphragms	Vacuum to a few hundred bars
Bellows	Vacuum to a hundred bar

Measurement of Vacuum

Pressures below atmosphere is vacuum. Very low pressures may be defined as below 1mmHg. Ultra low pressures is less than a milli micron ($<10^{-3}$ micron)

Measurement of vacuum may be by two methods

Direct measurement

> Resulting in a displacement caused by the action of force [Spiral Bourdon tubes, flat or corrugat ed diaphragm, capsules and various other manometers]

Indirect measurement or inferential methods

> Pressure is determined through the measurement of certain pressure controlled properties such as volume, thermal conductivity etc.

McLeod Gauge

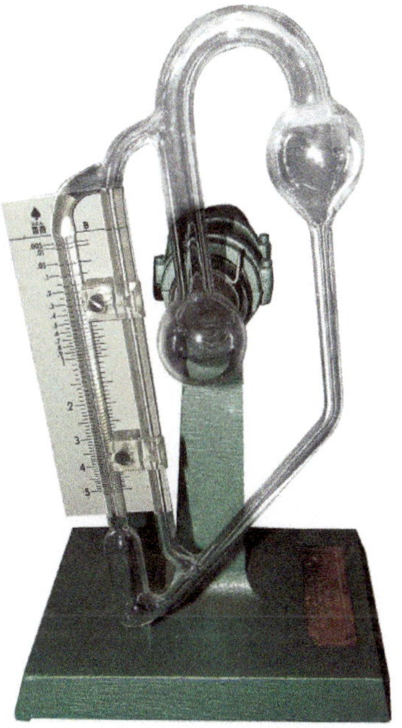

A glass McLeod gauge, drained of mercury

A McLeod gauge is a scientific instrument used to measure very low pressures, down to 10^{-6} Torr. It was invented in 1874 by Herbert McLeod (1841–1923). McLeod gauges were once commonly found attached to equipment that operates under vacuum, such as a lyophilizer. Today, however, these gauges have largely been replaced by electronic vacuum gauges.

The design of a McLeod gauge is somewhat similar to a that of a mercury-column manometer. Typically it is filled with mercury. If used incorrectly, this mercury can escape and contaminate the vacuum system attached to the gauge.

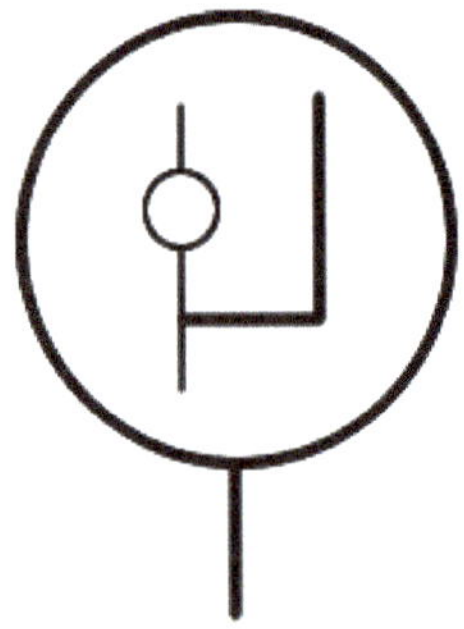

McLeod manometer symbol according to ISO 3753-1977 (E)

McLeod gauges operate by taking in a sample volume of gas from a vacuum chamber, then compressing it by tilting and infilling with mercury. The pressure in this smaller volume is then measured by a mercury manometer, and knowing the compression ratio (the ration of the initial and final volumes), the pressure of the original vacuum can be determined by applying Boyle's law.

This method is fairly accurate for non-condensible gases, such as oxygen and nitrogen. However, condensible gases, such as water vapour, ammonia, carbon dioxide, and pump-oil vapors may be in gaseous form in the low pressure of the vacuum chamber, but will condense when compressed by the McLeod gauge. The result is an erroneous reading, showing a pressure much lower than actually present. A cold trap may be used in conjunction with a McLeod gauge to condense these vapors before they enter the gauge.

The McLeod gauge has the advantage that it is simple to use and that its calibration is nearly the same for all non-condensable gases. The device can be manually operated and the scale read visually, or the process can be automated in various ways. For example, a small electric motor can periodically rotate the assembly to collect a gas sample. If a fine platinum wire is in the capillary tube, its resistance indicates the height of the mercury column around it.

Modern electronic vacuum gauges are simpler to use, less fragile, and do not present a mercury hazard, but their reading is highly dependent on the chemical nature of the gas being measured, and their calibration is unstable. For this reason, McLeod gauges continue to be used as a calibration standard for electronic gauges.

Inferential Methods

McLeod Gauge

The working of McLeod Gauge is based on Boyles' fundamental equation.

$$p_1 = \frac{p_2 V_2}{V_1}$$

where p and V refer to pressure and volume respectively and subscripts 1 and 2 refer to initial and final conditions. Conventional McLeod gauge is made of glass. Refer to the figure given below. It consists of the capillary 'C', bulb 'B' and the mercury sump which is connected to the lower end of the glass tube such that it can be moved up and down.

The pressure to be measured (the unknown pressure) is connected to the upper end of the glass part. When the mercury level in the gauge is below the cut off 'F', the unknown pressure fills the gauge including the bulb B and capillary C. When the mercury sump is moved up, the level in the gauge rises and when it reaches the cut off 'F' a known volume of gas at pressure to be measured is trapped in bulb B and capillary C.

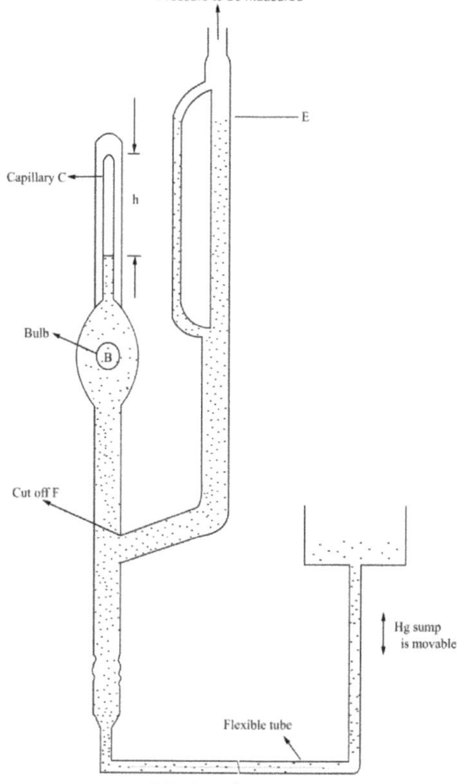

Mc Leod gauge

Mercury is then forced up into the bulb and capillary. Assume the sump is raised to such a level that the gas at the pressure to be measured which filled the volume above the cut off is now compressed to the volume represented by the column h.

Suppose the original volume after then mercury reaches F is V_O. This is at a pressure being measured p_1

$$V_0 \times p_1 = ah \times (p_1 + h)$$
$$= p_1\ ah + ah^2$$
$$V_0 p_1 - p_1\ ah = ah^2$$
$$p_1 (V_0 - ah) = ah^2$$

As 'ah' is <<< Vo it is neglected.

$$p_1 = \frac{ah^2}{V_0}$$

Applications of McLeod Gauge

McLeod gauge is used mainly for calibrating other inferential type of gauges. The shortcomings of the McLeod gauge are its fragility and the inability to measure continuously. The vapor pressure of Mercury sets the lower limit of measurement range of the gauge.

Thermal Conductivity Gauges

The working principle of thermal conductivity gauges is that at low pressures heat lost by a heated object by conduction through molecules will depend on pressure. This is valid only for certain pressure range.

When the mean free path is comparable with the dimensions of the gauge head the heat loss from a heated wire in the gauge head will be by (i) conduction through leads (ii) radiation to surroundings (iii) conduction through molecules.

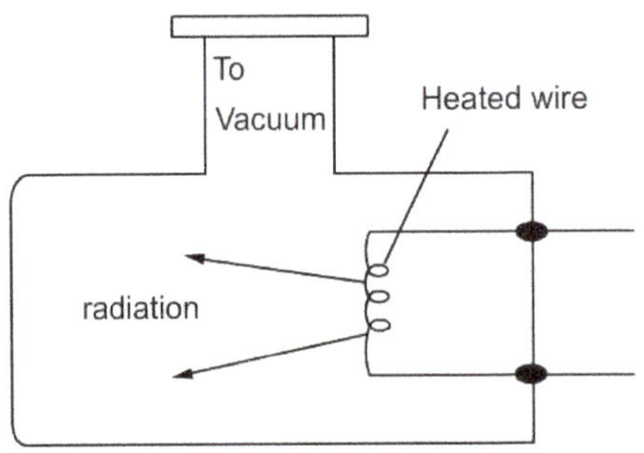

Gauge head of the thermal conductivity gauge

The range of thermal conductivity gauges is from mm of Hg to 10^{-3}mmHg. At higher pressures, heat loss from the heated wire is insensitive to the pressure change. At lower pressure heat loss by (i) and (ii) become more significant. There are two kinds of the thermocouple gauges.

Pirani Gauge

Measures change in resistance of the heated wire when it looses heat to the gas molecules in the gauge head. In this case the gauge is called pirani gauge.

Pirani Gauge

Pirani probe with 64 mm vacuum flange

The Pirani gauge is a robust thermal conductivity gauge used for the measurement of the pressures in vacuum systems. It was invented in 1906 by Marcello Pirani.

Structure

The Pirani gauge consists of a metal filament (usually platinum) suspended in a tube which is connected to the system whose vacuum is to be measured. Connection is usually made either by a ground glass joint or a flanged metal connector, sealed with an o-ring. The filament is connected to an electrical circuit from which, after calibration, a pressure reading may be taken.

Mode of Operation

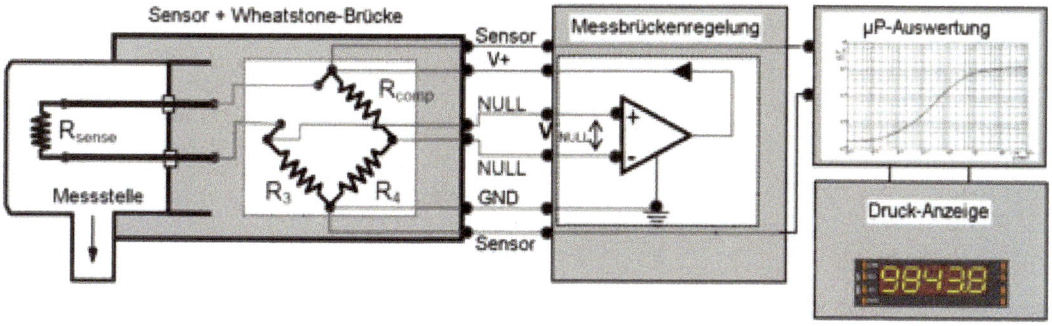

Block diagram of Pirani gauge

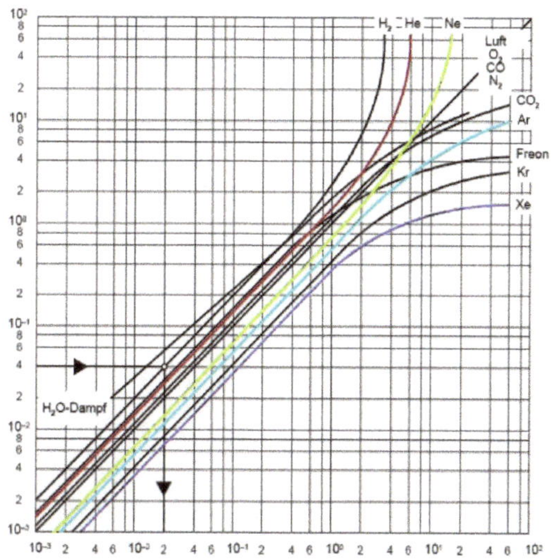

Curves to convert air readings to other gases

A heated metal wire (also called a filament) suspended in a gas will lose heat to the gas as its molecules collide with the wire and remove heat. If the gas pressure is reduced the number of molecules present will fall proportionately and the wire will lose heat more slowly. Measuring the heat loss is an indirect indication of pressure.

The electrical resistance of a wire varies with its temperature, so the resistance indicates the temperature of wire. In many systems, the wire is maintained at a constant resistance R by controlling the current I through the wire. The resistance can be set using a bridge circuit. The power delivered to the wire is I^2R, and the same power is transferred to the gas. The current required to achieve this balance is therefore a measure of the vacuum.

The gauge may be used for pressures between 0.5 Torr to 10^{-4} Torr. The thermal conductivity and heat capacity of the gas may affect the readout from the meter, and therefore the apparatus may need calibrating before accurate readings are obtainable. For lower pressure measurement, the thermal conductivity of the gas becomes increasingly smaller and more difficult to measure accurately, and other instruments such as a Penning gauge are used instead.

Pulsed Pirani Gauge

A special form of the Pirani gauge is the pulsed Pirani vacuum gauge where the filament is not operated at a constant temperature, but is cyclically heated up to a certain temperature threshold by an increasing voltage ramp. When the threshold is reached, the heating voltage is switched off and the filament cools down again. The required heat-up time is used as a measure of pressure.

For adequately low pressure the following relation for supplied heating power and filament temperature $T(t)$ applies:

$$P_{\text{el}} = C_1 \lambda_{\text{gas}} (T(t) - T_a) + C_2 \lambda_{\text{fil}} (T(t) - T_a) + A_{\text{fil}} \epsilon \sigma (T(t)^4 - T_a^4) + c_{\text{fil}} m_{\text{fil}} \frac{\mathrm{d}T}{\mathrm{d}t},$$

where c_{fil} is the heating capacity of the filament, m_{fil} is the mass of the filament and C_1 and C_2 are constants.

Advantages and Disadvantages of the Pulsed Gauge

Advantages

- Significantly better resolution in the range above 75 Torr.
- The power consumption is drastically reduced compared to continuously operated Pirani gauges.
- The gauge's thermal influence on the real measurement is lowered considerably due to the low temperature threshold of 80°C and the ramp heating in pulsed mode.
- The pulsed mode allows for efficient implementation of modern microprocessor technology.

Disadvantages

- Increased calibration effort
- Longer heat-up phase

Alternative

An alternative to the Pirani gauge is the thermocouple gauge, which works on the same principle of detecting thermal conductivity of the gas by a change in temperature. In the thermocouple gauge, the temperature is sensed by a thermocouple rather than by the change in resistance of the heated wire.

Thermocouple Gauge

Instead of measuring electrical resistance, a thermocouple is kept in contact with the heated wire and the temperature of the wire is directly measured as a measure of pressure.

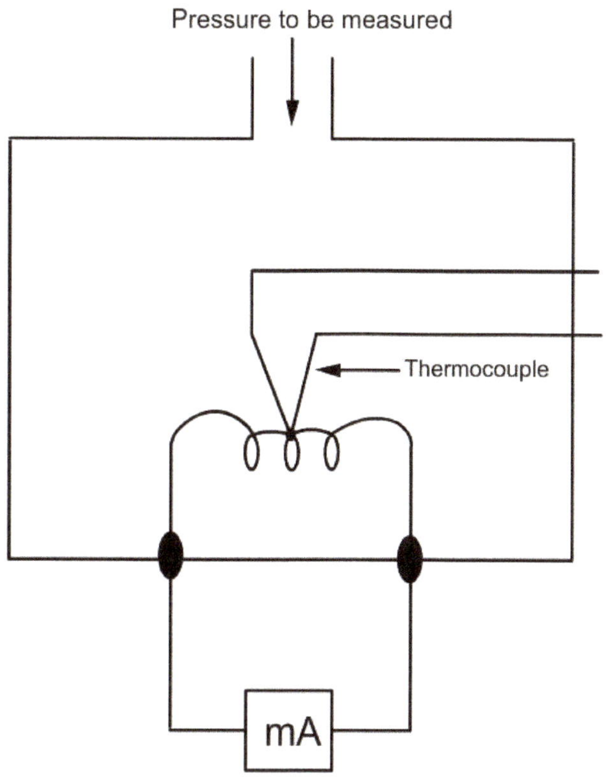

Thermocouple gauge head

Hot-filament Ionization Gauge

The hot-filament ionization gauge, sometimes called a hot-filament gauge or hot-cathode gauge, is the most widely used low-pressure (vacuum) measuring device for the region from 10^{-3} to 10^{-10} Torr. It is a triode, with the filament being the cathode.

Function

A regulated electron current (typically 10 mA) is emitted from a heated filament. The electrons are attracted to the helical grid by a DC potential of about +150 V. Most of the electrons pass through the grid and collide with gas molecules in the enclosed volume, causing a fraction of them to be ionized. The gas ions formed by the electron collisions are attracted to the central ion collector wire by the negative voltage on the collector (typically −30 V). Ion currents are on the order of 1 mA/Pa. This current is amplified and displayed by a high-gain differential amplifier/electrometer.

This ion current differs for different gases at the same pressure; that is, a hot-filament ionization gauge is composition-dependent. Over a wide range of molecular density, however, the ion current from a gas of constant composition is directly proportional to the molecular density of the gas in the gauge.

Construction

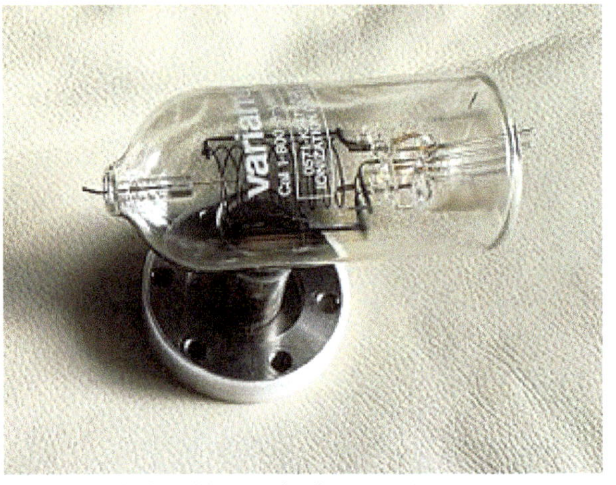

Tubulated hot-cathode ionization gauge

A hot-cathode ionization gauge is composed mainly of three electrodes, all acting as a triode, wherein the cathode is the filament. The three electrodes are a collector or plate, a filament, and a grid. The collector current is measured in picoamperes by an electrometer. The filament voltage to ground is usually at a potential of 30 volts, while the grid voltage at 180–210 volts DC, unless there is an optional electron bombardment feature, by heating the grid, which may have a high potential of approximately 565 volts. The most common ion gauge is the hot-cathode Bayard–Alpert gauge, with a small collector inside the grid. A glass envelope with an opening to the vacuum can surround the electrodes, but usually the nude gauge is inserted in the vacuum chamber directly, the pins being fed through a ceramic plate in the wall of the chamber. Hot-cathode gauges can be damaged or lose their calibration if they are exposed to atmospheric pressure or even low vacuum while hot.

Electrons emitted from the filament move several times in back-and-forth movements around the grid before finally entering the grid. During these movements, some electrons collide with a gas molecule to form a pair of an ion and an electron (electron ionization). The number of these ions is proportional to the gas molecule density multiplied by the electron current emitted from the filament, and these ions pour into the collector to form an ion current. Since the gas molecule density is proportional to the pressure, the pressure is estimated by measuring the ion current.

The low-pressure sensitivity of hot-cathode gauges is limited by the photoelectric effect. Electrons hitting the grid produce X-rays that produce photoelectric noise in the ion collector. This limits the range of older hot-cathode gauges to 10^{-8} Torr and the Bayard–Alpert gauges to about 10^{-10} Torr. Additional wires at cathode potential in the line of sight between the ion collector and the grid prevent this effect. In the extraction type the ions are not attracted by a wire but by an open cone. As the ions cannot decide which part of the cone to hit, they pass through the hole and form an ion beam. This ion beam can be passed on to a:

- Faraday cup,
- quadrupole mass analyzer with Faraday cup,
- microchannel plate detector with Faraday cup,
- quadrupole mass analyzer with microchannel plate detector Faraday cup,
- ion lens and acceleration voltage and directed at a target to form a sputter gun; in this case a valve lets gas into the grid cage.

Types of Hot-filament Ionization Gauges

- Bayard–Alpert (uses sealed tube).
- Nude gauge (uses the vacuum chamber to make a complete seal).

Hot Cathode Ionisation Gauges

At higher levels of vacuum, the measurement of pressure is by using Ionisation gauges. They operate on the principle of ionising the gas by means of electrons emitted by a heated filament. The kinetic energy acquired by an electron in passing through a potential difference of V volts corresponds to a value equal to V*e where e is the charge of the electron. When this energy exceeds a certain critical value corresponding to the ionization potential Vi, there is a possibility that collisions between molecules and electrons will result in the formation of +ve ions. The relatively high velocity electrons on hitting a gas molecule drives an electron out of it leaving it positively charged. For gases such as N_2, O_2 etc, Vi is ~ 15volts. The measurement of the ions produced is a measure of the pressure. The electrons are speeded up by an electric field and +ve ions produced are collected. The number of +ve ions formed will depend on the number of molecules and therefore on the pressure.

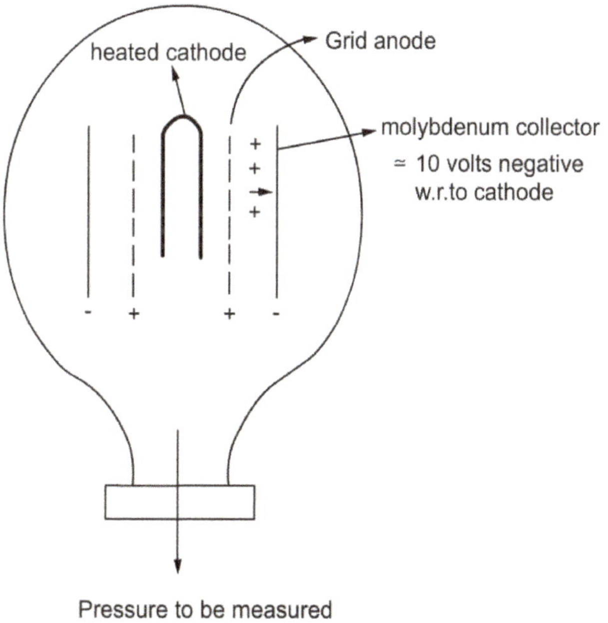

Ionisation gauge

The gauge consists of a cathode, grid anode and a negative plate. The negative plate is at ~10V negative with respect to the cathode. The electrons emitted by the hot cathode (filament) are speeded up by the electric field and the positive ions produced are collected by the negative plate.

Cold Cathode Ionisation Gauge (Penning Gauge)

This gauge also works on the ionization principle. Positive ions are produced by the electrons and current due to these ions gives a measure of the pressure. Electrons are ejected from a cold cathode of Zirconium, Thorium by electric discharge. The gauge consists of two plate cathodes and a ring anode . A potential difference of ~2KV is applied across the electrodes.

The travel of electrons is made over a much longer distance. The secondary electrons are made to travel in helical paths before reaching the anode. This is accomplished by a magnetic field.

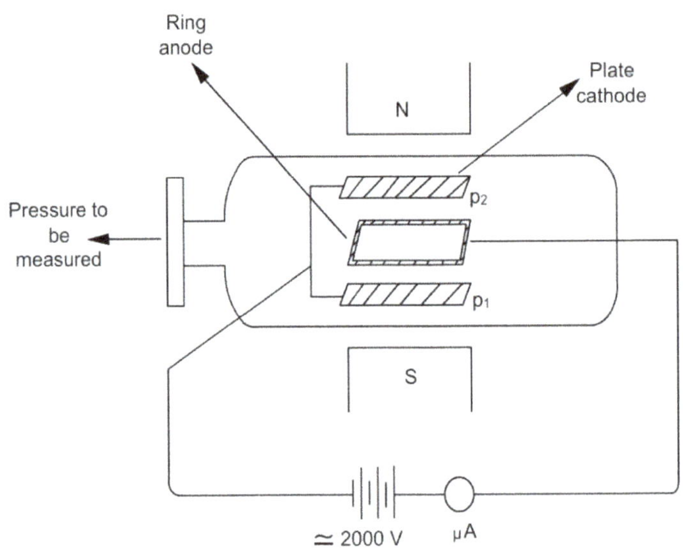

Cold cathode ionization gauge

Magnetic poles are kept such that the flux with lines of force is applied perpendicular to the two cathodes.

Measurement of Pressure in Flows

In flows, distinction has to be made between static and stagnation pressures.

Static Pressure

Pressure acting on the surface of a body imagined to be moving with the fluid with the same velocity as the medium is the static pressure.

Stagnation Pressure

It is the pressure of a fluid imagined to be brought to rest isentropically.

Measurement of Static Pressure

Common technique is to connect a probe to an orifice drilled perpendicular to the wall of the model where the streamlines are undistorted and parallel.

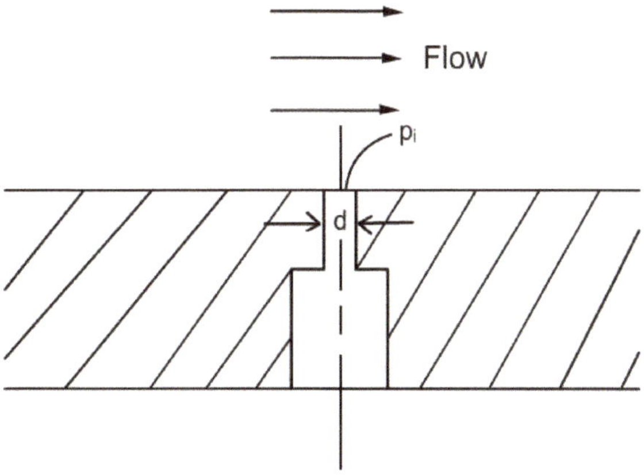

Static pressure orifice (tap) on a flat wall

Static hole diameter is about 1/5 of the boundary layer thickness. Practically the diameter is about 0.25mm on small models and 2.5mm on larger installations. The correctness of the static pressure measured is dependent on 'd' the orifice diameter. The influence of the orifice diameter on the error in static pressure measurements is given graphically in the figure given below.

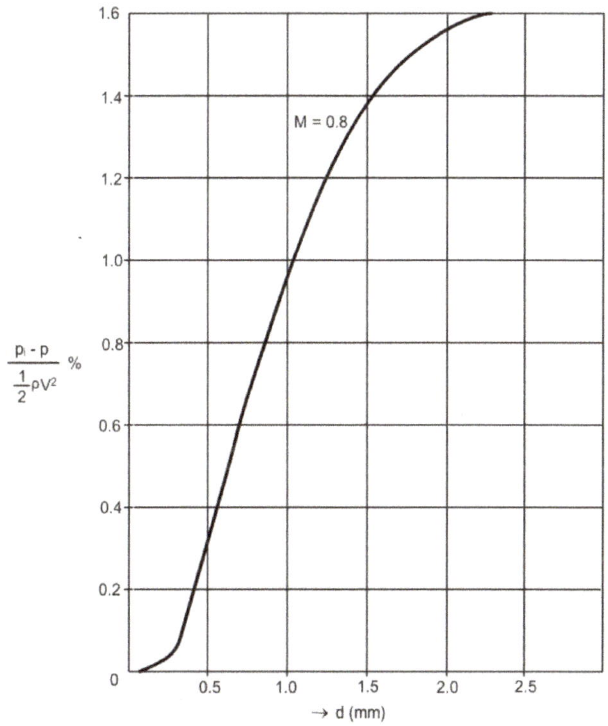

Influence of error on orifice diameter

The static pressure orifices (taps) on the walls of the flow channel provide the wall static pressure. The wall static pressure can not be assumed to prevail inside the flow. In order to get the pressure inside the flow, probes of suitable design have to be made use of.

Static Pressure Probes for Subsonic Flow

The commonly used static pressure probe in subsonic flows is the Prandtl probe. The Prandtl probe is an intrusive device. The pressure sensing orifices on the periphery of the probe are carefully located such that the influence of the nose and stem of the probe nullify each other.

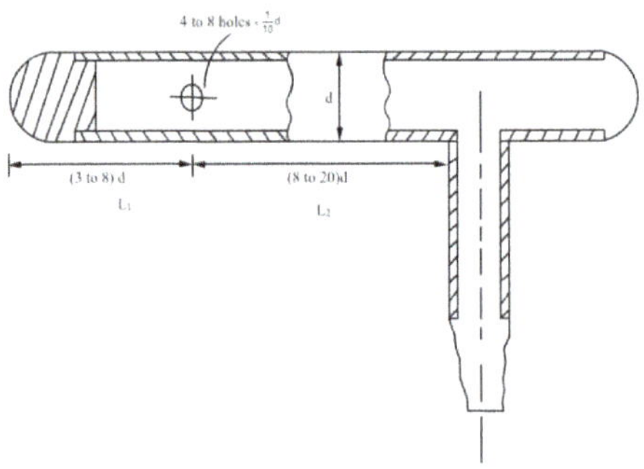

(a) The Prandtl probe

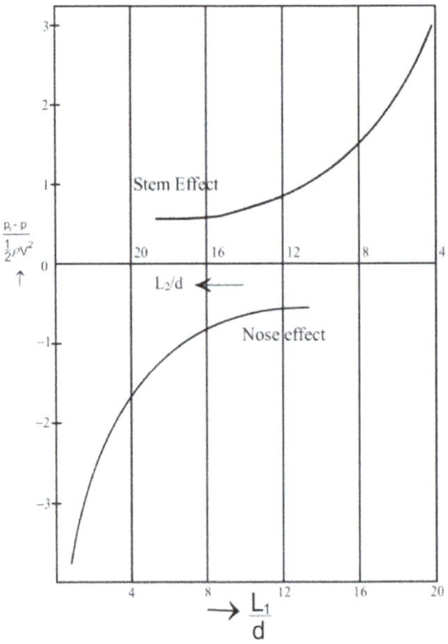

(b) The nose and stem effects in a Prandtl tube

Because of the intrusive nature of the probe, the flow will be accelerated by the nose. The effect will be to reduce the static pressure which is called the nose effect. In contrast, the effect of stem will be to locally stagnate the flow and thereby to increase the static pressure. This effect is the stem effect. As shown in the fig. (b), the position of the orifices are so chosen that the two effects neutralize each other. Additionally the probe must be slender (say ~ 1.0 to 1.5mm dia) and kept parallel to the flow.

Static Pressure Probes for Supersonic Flow

When Mach number is more than 1.0, shock appears. When the cone angle of the probe is less than the shock detachment angle for the given Mach number shown in fig. (c) and if the orifices are located well downstream of the shock wave, the measured static pressure will tend towards the value for the undisturbed flow. Conical or ogival shaped tubes are used.

Probes are made small and with static taps on the cone surface. The effect of yaw is reduced by arranging several orifices so that the pressure inside the tube is an average value. Usually the tube has 4 to 8 orifices whose diameter is about 1/10 th of the outside diameter of the tube.

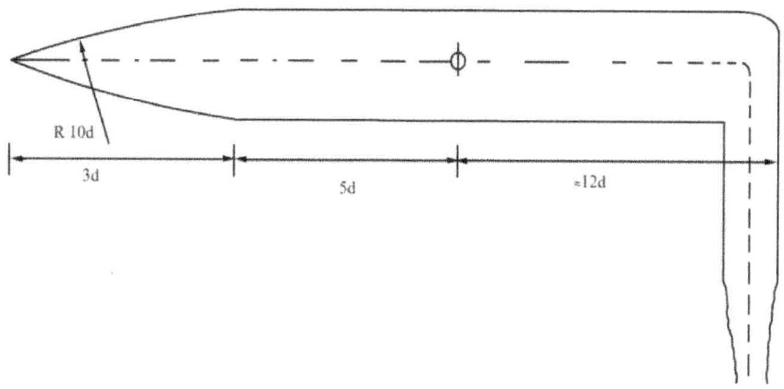

(a) Ogival shaped static pressure probe for supersonic flow

Good results are obtained with ogival tubes. The tube shown in figure above has a systematic error within 1%. Angle of the cone should be less than the angle at which the shock wave becomes detached from the cone. The error in the measurement of static pressure decreases as the distance of the orifice from the tip of the probe increases.

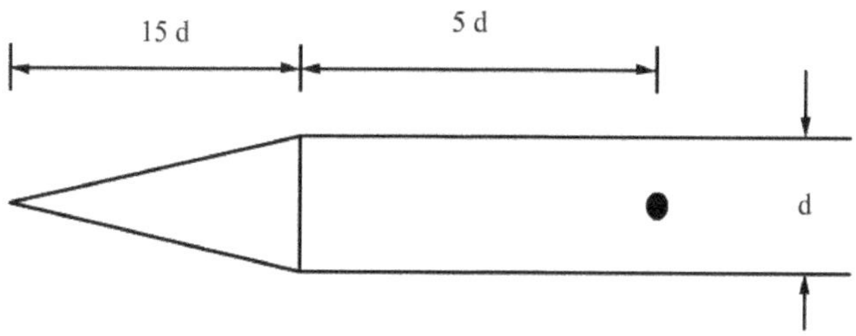

(b) Conical static pressure probe for supersonic flow

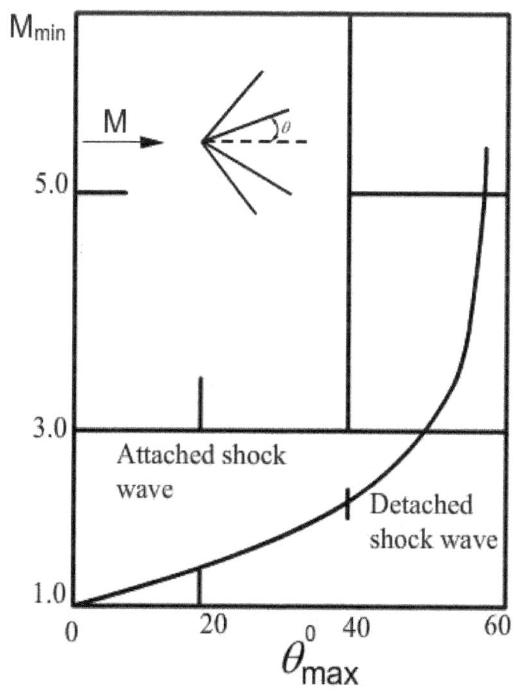

(c) Maximum deflection angle for different Mach numbers

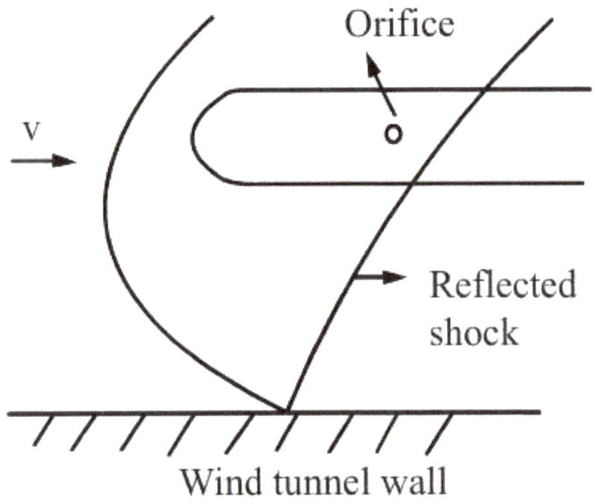

(d) Reflection of the shock from the wall

It is important to have long pointed tubes. Otherwise there is the possibility that the reflected shock may affect the static pressure readings. Roughness at the edges of the orifices too may cause large errors in measurement.

Measurement of Stagnation Or Total Pressure

Insubsonic Flows

The gas particles come to rest so quickly at the stagnation point of a body, that heat transfer and friction losses are negligible. Therefore, in subsonic flows only isentropic changes occur. Hence the stagnation pressure measured in a subsonic flow is not significantly different from that in the settling chamber. Total pressure is measured with a Pitot tube which a cylindrical tube is having an orifice pointing towards the flow. Axis of tube must coincide with the flow direction. It is a practice to use tubes which generally have blunt ends.

Such tubes are insensitive to yaw upto ± 10 to 12°

The relation between the stagnation pressure measured and the static pressure may be expressed as follows:

$$\frac{p}{p_0} = \left(1 + \frac{r-1}{2} M^2\right)^{\frac{r}{r-1}}$$

where p_0 is the total pressure and p the static pressure.

Insupersonic Velocities

Shockwave appears upstream of the tube nose. Therefore tube measures only pressure behind the shock wave. Normal shock equations give the relation between the total pressures upstream and downstream of the shock. As the bow shock formed in front of the Pitot tube has the normal part only in the central region, the tube diameter for measurements in supersonic flow is usually kept very small.

Lag in Manometric Systems

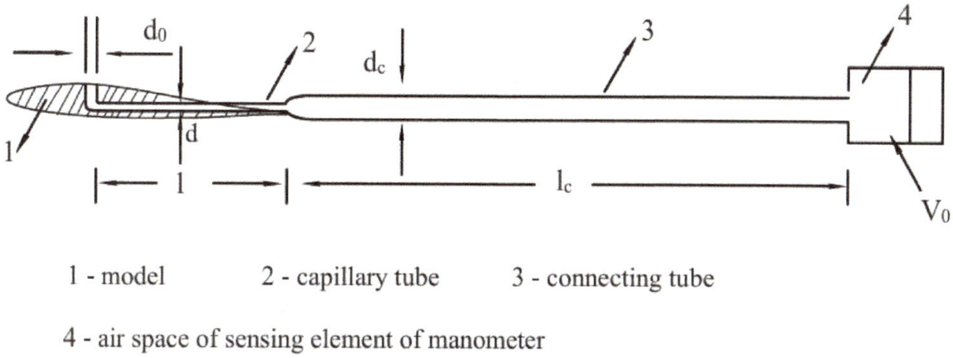

Wind tunnel model with the manometric system

When the pressure changes near the orifice which is connected to a manometer, equilibrium in the manometer is established not immediately but after a certain time. If the pressure is read of earlier there will be gross errors.

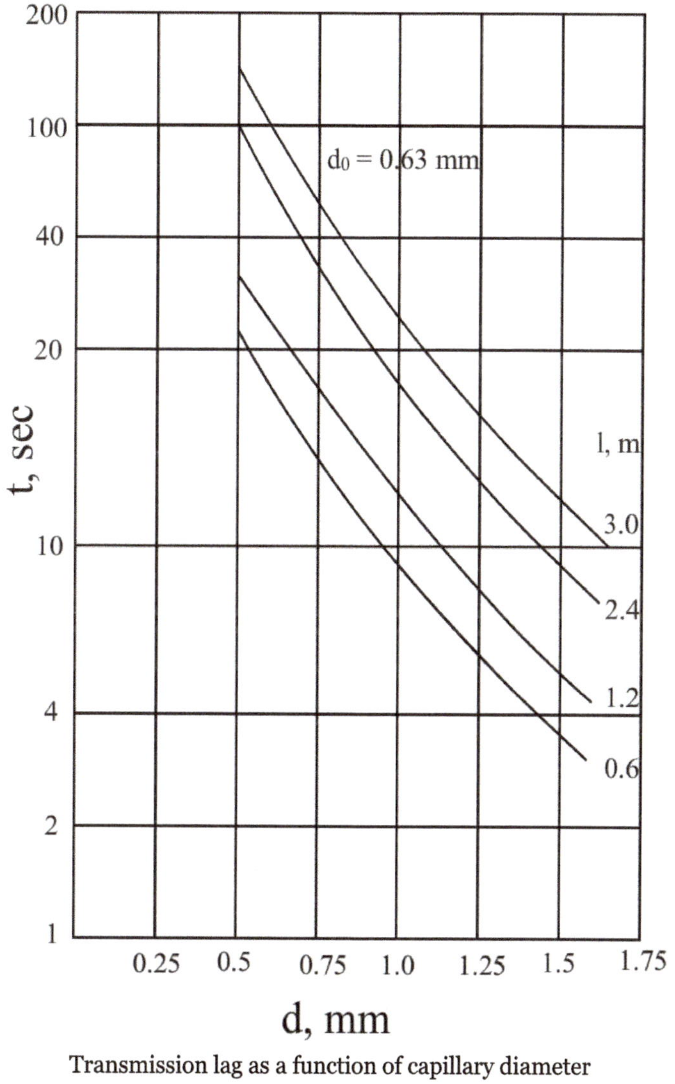

Transmission lag as a function of capillary diameter

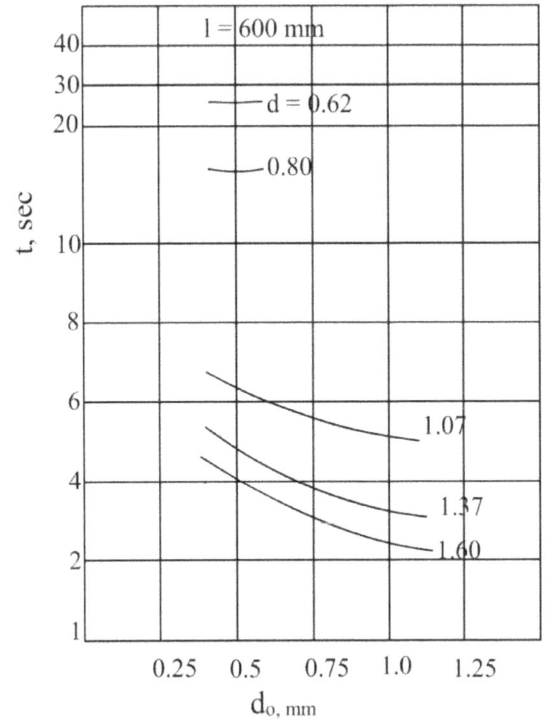

Transmission lag as a function of orifice diameter

Smaller transmission lags are necessary not only for high reliability but also for reducing the duration of experiments. Equilibrium will be established in the manometric system later after the pressure on the model is stabilized. The run times should be longer than the transmission lag.

Transmission lag is caused by:

a) Resistance of the tubes

b) The change in air density

c) and inertia of the moving masses

The main factors influencing the transmission lag are the orifice diameter d_o, the diameter d of the capillary and d_c of the connecting tube and their respective lengths l and l_c .

The orifice diameter is of small influence when $d/d_o < 2.5$. When $d/d_o > 2.5$ the transmission lag increases sharply. The orifice diameter should not be less than half the diameter of the capillary tube. The influence of the diameter of the capillary tube is very strong. A reduction of this diameter has its main effect an increase in the resistance to the flow of gas. An increase in the length of the capillary tube has a significant effect on the lag. Capillary tubes should have larger diameter and shorter length.

4

Temperature and its Measurement

Temperature can be measured by several scales and units. One of the devices used for temperature sensors is thermistor. Laws of thermocouples are an important part of the following section. The aspects elucidated in this section are of vital importance, and provide a better understanding of temperature measurement.

Temperature

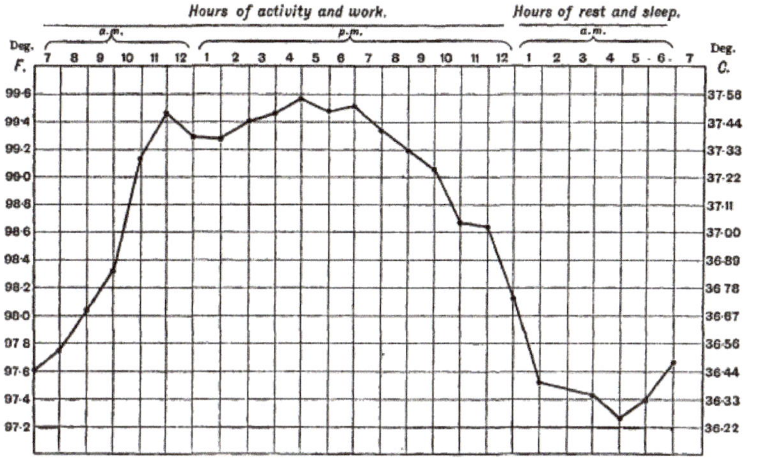

Body temperature variation

A temperature is an objective comparative measurement of hot or cold. It is measured by a thermometer. Several scales and units exist for measuring temperature, the most common being Celsius (denoted °C; formerly called *centigrade*), Fahrenheit (denoted °F), and, especially in science, Kelvin (denoted K).

The coldest theoretical temperature is absolute zero, at which the thermal motion of atoms and molecules reaches its minimum – classically, this would be a state of motionlessness, but quantum uncertainty dictates that the particles still possess a finite zero-point energy. Absolute zero is denoted as 0 K on the Kelvin scale, −273.15 °C on the Celsius scale, and −459.67 °F on the Fahrenheit scale.

The kinetic theory offers a valuable but limited account of the behavior of the materials of macroscopic bodies, especially of fluids. It indicates the absolute temperature as proportional to the average kinetic energy of the random microscopic motions of those of their constituent microscopic particles, such as electrons, atoms, and molecules, that move freely within the material.

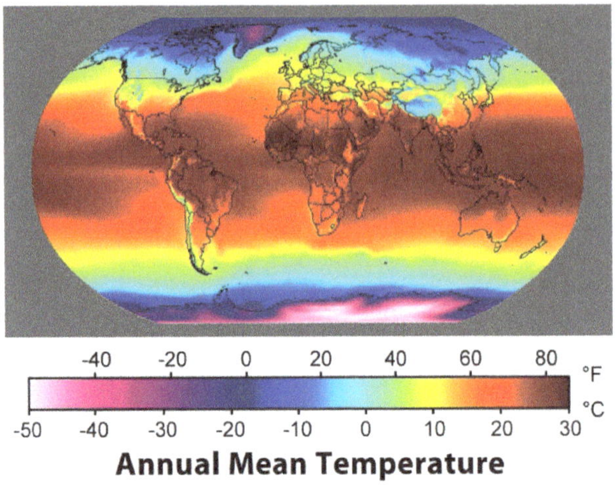

Annual mean temperature around the world

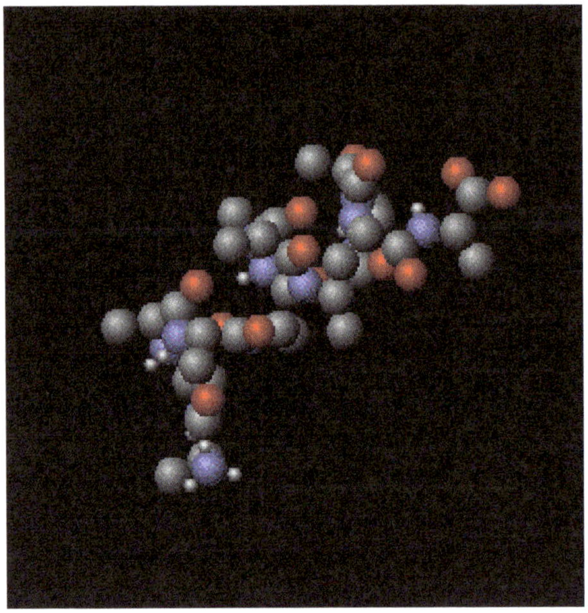

Thermal vibration of a segment of protein alpha helix: The amplitude of the vibrations increases with temperature.

Temperature is important in all fields of natural science including physics, geology, chemistry, atmospheric sciences, medicine and biology as well as most aspects of daily life.

Effects of Temperature

Many physical processes are affected by temperature, such as

- physical properties of materials including the phase (solid, liquid, gaseous or plasma), density, solubility, vapor pressure, electrical conductivity
- rate and extent to which chemical reactions occur
- the amount and properties of thermal radiation emitted from the surface of an object
- speed of sound is a function of the square root of the absolute temperature

Temperature Scales

Temperature scales differ in two ways: the point chosen as zero degrees, and the magnitudes of incremental units or degrees on the scale.

The Celsius scale (°C) is used for common temperature measurements in most of the world. It is an empirical scale. It developed by a historical progress, which led to its zero point 0°C being defined by the freezing point of water, with additional degrees defined so that 100°C was the boiling point of water, both at sea-level atmospheric pressure. Because of the 100 degree interval, it is called a centigrade scale. Since the standardization of the kelvin in the International System of Units, it has subsequently been redefined in terms of the equivalent fixing points on the Kelvin scale, and so that a temperature increment of one degree Celsius is the same as an increment of one kelvin, though they differ by an additive offset of 273.15.

The United States commonly uses the Fahrenheit scale, on which water freezes at 32°F and boils at 212°F at sea-level atmospheric pressure.

Many scientific measurements use the Kelvin temperature scale (unit symbol: K), named in honor of the Scottish physicist who first defined it. It is a thermodynamic or absolute temperature scale. Its zero point, 0K, is defined to coincide with the coldest physically-possible temperature (called absolute zero). Its degrees are defined through thermodynamics. The temperature of absolute zero occurs at 0K = −273.15°C (or −459.67°F), and the freezing point of water at sea-level atmospheric pressure occurs at 273.15K = 0°C.

The International System of Units (SI) defines a scale and unit for the kelvin or thermodynamic temperature by using the reliably reproducible temperature of the triple point of water as a second reference point (the first reference point being 0 K at absolute zero). The triple point is a singular state with its own unique and invariant temperature and pressure, along with, for a fixed mass of water in a vessel of fixed volume, an autonomically and stably self-determining partition into three mutually contacting phases, vapour, liquid, and solid, dynamically depending only on the total internal energy of the mass of water. For historical reasons, the triple point temperature of water is fixed at 273.16 units of the measurement increment.

Thermodynamic Approach to Temperature

Temperature is one of the principal quantities in the study of thermodynamics.

Kinds of Temperature Scale

There is a variety of kinds of temperature scale. It may be convenient to classify them as empirically and theoretically based. Empirical temperature scales are historically older, while theoretically based scales arose in the middle of the nineteenth century.

Empirically Based Scales

Empirically based temperature scales rely directly on measurements of simple physical properties of materials. For example, the length of a column of mercury, confined in a glass-walled capillary tube, is dependent largely on temperature, and is the basis of the very useful mercury-in-glass thermometer. Such scales are valid only within convenient ranges of temperature. For example, above the boiling point of mercury, a mercury-in-glass thermometer is impracticable. Most materials expand with temperature increase, but some materials, such as water, contract with temperature increase over some specific range, and then they are hardly useful as thermometric materials. A material is of no use as a thermometer near one of its phase-change temperatures, for example its boiling-point.

In spite of these restrictions, most generally used practical thermometers are of the empirically based kind. Especially, it was used for calorimetry, which contributed greatly to the discovery of thermodynamics. Nevertheless, empirical thermometry has serious drawbacks when judged as a basis for theoretical physics. Empirically based thermometers, beyond their base as simple direct measurements of ordinary physical properties of thermometric materials, can be re-calibrated, by use of theoretical physical reasoning, and this can extend their range of adequacy.

Theoretically Based Scales

Theoretically based temperature scales are based directly on theoretical arguments, especially those of thermodynamics, kinetic theory and quantum mechanics. They rely on theoretical properties of idealized devices and materials. They are more or less comparable with practically feasible physical devices and materials. Theoretically based temperature scales are used to provide calibrating standards for practical empirically based thermometers.

The accepted fundamental thermodynamic temperature scale is the Kelvin scale, based on an ideal cyclic process envisaged for a Carnot heat engine.

An ideal material on which a temperature scale can be based is the ideal gas. The pressure exerted by a fixed volume and mass of an ideal gas is directly proportional to its temperature. Some natural gases show so nearly ideal properties over suitable temperature ranges that they can be used for thermometry; this was important during the development of thermodynamics and is still of practical importance today. The ideal gas thermometer is, however, not theoretically perfect for thermodynamics. This is because the entropy of an ideal gas at its absolute zero of temperature is not a positive semi-definite quantity, which puts the gas in violation of the third law of thermodynamics. The physical reason is that the ideal gas law, exactly read, refers to the limit of infinitely high temperature and zero pressure.

Measurement of the spectrum of electromagnetic radiation from an ideal three-dimensional black body can provide an accurate temperature measurement because the frequency of maximum spectral radiance of black-body radiation is directly proportional to the temperature of the black body; this is known as Wien's displacement law and has a theoretical explanation in Planck's law and the Bose–Einstein law.

Measurement of the spectrum of noise-power produced by an electrical resistor can also provide an accurate temperature measurement. The resistor has two terminals and is in effect a one-dimensional body. The Bose-Einstein law for this case indicates that the noise-power is directly proportional to the temperature of the resistor and to the value of its resistance and to the noise band-width. In a given frequency band, the noise-power has equal contributions from every frequency and is called Johnson noise. If the value of the resistance is known then the temperature can be found.

If molecules, or atoms, or electrons, are emitted from a material and their velocities are measured, the spectrum of their velocities often nearly obeys a theoretical law called the Maxwell–Boltzmann distribution, which gives a well-founded measurement of temperatures for which the law holds. There have not yet been successful experiments of this same kind that directly use the Fermi–Dirac distribution for thermometry, but perhaps that will be achieved in future.

Absolute Thermodynamic Scale

The Kelvin scale is called absolute for two reasons. One is Kelvin's, that its formal character is independent of the properties of particular materials. The other reason is that its zero is in a sense absolute, in that it indicates absence of microscopic classical motion of the constituent particles of matter, so that they have a limiting specific heat of zero for zero temperature, according to the third law of thermodynamics. Nevertheless, a Kelvin temperature does in fact have a definite numerical value that has been arbitrarily chosen by tradition and is dependent on the property of a particular materials; it is simply less arbitrary than relative "degrees" scales such as Celsius and Fahrenheit. Being an absolute scale with one fixed point (zero), there is only one degree of freedom left to arbitrary choice, rather than two as in relative scales. For the Kelvin scale in modern times, this choice of convention is made to be that of setting the gas–liquid–solid triple point of water, a point which can be reliably reproduced as a standard experimental phenomenon, at a numerical value of 273.16 kelvins. The Kelvin scale is also called the thermodynamic scale. However, to demonstrate that its numerical value is indeed arbitrary, it is useful to point out that an alternate, less widely used absolute temperature scale exists called the Rankine scale, made to be aligned with the Fahrenheit scale as Kelvin is with Celsius.

Definition of the Kelvin Scale

The thermodynamic definition of temperature is due to Kelvin.

It is framed in terms of an idealized device called a Carnot engine, imagined to define a continuous cycle of states of its working body. The cycle is imagined to run so slowly that at each point of the cycle the working body is in a state of thermodynamic equilibrium. There are four limbs in such a Carnot cycle. The engine consists of four bodies. The main one is called the working body. Two of them are called heat reservoirs, so large that their respective non-deformation variables are not changed by transfer of energy as heat through a wall permeable only to heat to the working body. The fourth body is able to exchange energy with the working body only through adiabatic work; it may be called the work reservoir. The substances and states of the two heat reservoirs should be chosen so that they are not in thermal equilibrium with one another. This means that they must be at different fixed temperatures, one, labeled here with the number 1, hotter than the other, labeled here with the number 2. This can be tested by connecting the heat reservoirs successively to an auxiliary empirical thermometric body that starts each time at a convenient fixed intermediate temperature. The thermometric body should be composed of a material that has a strictly monotonic relation between its chosen empirical thermometric variable and the amount of adiabatic isochoric work done on it. In order to settle the structure and sense of operation of the Carnot cycle, it is convenient to use such a material also for the working body; because most materials are of this kind, this is hardly a restriction of the generality of this definition. The Carnot cycle is considered to start from an initial condition of the working body that was reached by the completion of a reversible adiabatic compression. From there, the working body is initially connected by a wall permeable only to heat to the heat reservoir number 1, so that during the first limb of the cycle it expands and does work on the work reservoir. The second limb of the cycle sees the working body expand adiabatically and reversibly, with no energy exchanged as heat, but more energy being transferred as work to the work reservoir. The third limb of the cycle sees the working body connected, through a wall permeable only to heat, to the heat reservoir 2, contracting and accepting energy as work from the work reservoir. The cycle is closed by reversible adiabatic compression of the working body, with no energy transferred as heat, but energy being transferred to it as work from the work reservoir.

With this set-up, the four limbs of the reversible Carnot cycle are characterized by amounts of energy transferred, as work from the working body to the work reservoir, and as heat from the heat reservoirs to the working body. The amounts of energy transferred as heat from the heat reservoirs are measured through the changes in the non-deformation variable of the working body, with reference to the previously known properties of that body, the amounts of work done on the work reservoir, and the first law of thermodynamics. The amounts of energy transferred as heat respectively from reservoir 1 and from reservoir 2 may then be denoted respectively Q_1 and Q_2. Then the absolute or thermodynamic temperatures, T_1 and T_2, of the reservoirs are defined so that to be such that

$$T_1 / T_2 = -Q_1 / Q_2 \, . \tag{1}$$

Kelvin's original work postulating absolute temperature was published in 1848. It was based on the work of Carnot, before the formulation of the first law of thermodynamics. Kelvin wrote in his 1848 paper that his scale was absolute in the sense that it was defined "independently of the properties of any particular kind of matter." His definitive publication, which sets out the definition just stated, was printed in 1853, a paper read in 1851.

This definition rests on the physical assumption that there are readily available walls permeable only to heat. In his detailed definition of a wall permeable only to heat, Carathéodory includes several ideas. The non-deformation state variable of a closed system is represented as a real number. A state of thermal equilibrium between two closed systems connected by a wall permeable only to heat means that a certain mathematical relation holds between the state variables, including the respective non-deformation variables, of those two systems (that particular mathematical relation is regarded by Buchdahl as a preferred statement of the zeroth law of thermodynamics). Also, referring to thermal contact equilibrium, "whenever each of the systems S_1 and S_2 is made to reach equilibrium with a third system S_3 under identical conditions, the systems S_1 and S_2 are in mutual equilibrium." It may be viewed as a re-statement of the principle stated by Maxwell in the words: "All heat is of the same kind." This physical idea is also expressed by Bailyn as a possible version of the zeroth law of thermodynamics: "All diathermal walls are equivalent." Thus the present definition of thermodynamic temperature rests on the zeroth law of thermodynamics. Explicitly, this present definition of thermodynamic temperature also rests on the first law of thermodynamics, for the determination of amounts of energy transferred as heat.

Implicitly for this definition, the second law of thermodynamics provides information that establishes the virtuous character of the temperature so defined. It provides that any working substance that complies with the requirement stated in this definition will lead to the same ratio of thermodynamic temperatures, which in this sense is universal, or absolute. The second law of thermodynamics also provides that the thermodynamic temperature defined in this way is positive, because this definition requires that the heat reservoirs not be in thermal equilibrium with one another, and the cycle can be imagined to operate only in one sense if net work is to be supplied to the work reservoir.

Numerical details are settled by making one of the heat reservoirs a cell at the triple point of water, which is defined to have an absolute temperature of 273.16 K. The zeroth law of thermodynamics allows this definition to be used to measure the absolute or thermodynamic temperature of an arbitrary body of interest, by making the other heat reservoir have the same temperature as the body of interest.

Temperature as an Intensive Variable

In thermodynamic terms, temperature is an intensive variable because it is equal to a differential coefficient of one extensive variable with respect to another, for a given body. It thus has the dimensions of a ratio of two extensive variables. In thermodynamics, two bodies are often considered as connected by contact with a common wall, which has some specific permeability properties. Such specific permeability can be referred to a specific intensive variable. An example is a diathermic wall that is permeable only to heat; the intensive variable for this case is temperature. When the two bodies have been in contact for a very long time, and have settled to a permanent steady state, the relevant intensive variables are equal in the two bodies; for a diathermal wall, this statement is sometimes called the zeroth law of thermodynamics.

In particular, when the body is described by stating its internal energy U, an extensive variable, as a function of its entropy S, also an extensive variable, and other state variables V, N, with $U = U(S, V, N)$, then the temperature is equal to the partial derivative of the internal energy with respect to the entropy:

$$T = \left(\frac{\partial U}{\partial S}\right)_{V,N}. \qquad (2)$$

Likewise, when the body is described by stating its entropy S as a function of its internal energy U, and other state variables V, N, with $S = S(U, V, N)$, then the reciprocal of the temperature is equal to the partial derivative of the entropy with respect to the internal energy:

$$\frac{1}{T} = \left(\frac{\partial S}{\partial U}\right)_{V,N}. \qquad (3)$$

The above definition, equation (1), of the absolute temperature is due to Kelvin. It refers to systems closed to transfer of matter, and has special emphasis on directly experimental procedures. A presentation of thermodynamics by Gibbs starts at a more abstract level and deals with systems open to the transfer of matter; in this development of thermodynamics, the equations (2) and (3) above are actually alternative definitions of temperature.

Temperature Local when Local Thermodynamic Equilibrium Prevails

Real world bodies are often not in thermodynamic equilibrium and not homogeneous. For study by methods of classical irreversible thermodynamics, a body is usually spatially and temporally divided conceptually into 'cells' of small size. If classical thermodynamic equilibrium conditions for matter are fulfilled to good approximation in such a 'cell', then it is homogeneous and a temperature exists for it. If this is so for every 'cell' of the body, then local thermodynamic equilibrium is said to prevail throughout the body.

It makes good sense, for example, to say of the extensive variable U, or of the extensive variable S, that it has a density per unit volume, or a quantity per unit mass of the system, but it makes no sense to speak of density of temperature per unit volume or quantity of temperature per unit mass of the system. On the other hand, it makes no sense to speak of the internal energy at a point, while when local thermodynamic equilibrium prevails, it makes good sense to speak of the temperature at a point. Consequently, temperature can vary from point to point in a medium that is not in global thermodynamic equilibrium, but in which there is local thermodynamic equilibrium.

Thus, when local thermodynamic equilibrium prevails in a body, temperature can be regarded as a spatially varying local property in that body, and this is because temperature is an intensive variable.

Kinetic Theory Approach to Temperature

A more thorough account of this is below at Theoretical foundation.

Kinetic theory provides a microscopic explanation of temperature, based on macroscopic systems' being composed of many microscopic particles, such as molecules and ions of various species, the particles of a species being all alike. It explains macroscopic phenomena through the classical mechanics of the microscopic particles. The equipartition theorem of kinetic theory asserts that each classical degree of freedom of a freely moving particle has an average kinetic energy of $k_B T/2$ where k_B denotes Boltzmann's constant. The translational motion of the particle has three degrees of freedom, so that, except at very low temperatures where quantum effects predominate, the average translational kinetic energy of a freely moving particle in a system with temperature T will be $3k_B T/2$.

It is possible to measure the average kinetic energy of constituent microscopic particles if they are allowed to escape from the bulk of the system. The spectrum of velocities has to be measured, and the average calculated from that. It is not necessarily the case that the particles that escape and are measured have the same velocity distribution as the particles that remain in the bulk of the system, but sometimes a good sample is possible.

Molecules, such as oxygen (O_2), have more degrees of freedom than single spherical atoms: they undergo rotational and vibrational motions as well as translations. Heating results in an increase in temperature due to an increase in the average translational kinetic energy of the molecules. Heating will also cause, through equipartitioning, the energy associated with vibrational and rotational modes to increase. Thus a diatomic gas will require more energy input to increase its temperature by a certain amount, i.e. it will have a greater heat capacity than a monatomic gas.

The process of cooling involves removing internal energy from a system. When no more energy can be removed, the system is at absolute zero, though this cannot be achieved experimentally. Absolute zero is the null point of the thermodynamic temperature scale, also called absolute temperature. If it were possible to cool a system to absolute zero, all classical motion of its particles would cease and they would be at complete rest in this classical sense. Microscopically in the description of quantum mechanics, however, matter still has zero-point energy even at absolute zero, because of the uncertainty principle.

Basic Theory

Temperature is a measure of a quality of a state of a material The quality may be regarded as a more abstract entity than any particular temperature scale that measures it, and is called *hotness* by some writers. The quality of hotness refers to the state of material only in a particular locality, and in general, apart from bodies held in a steady state of thermodynamic equilibrium, hotness varies from place to place. It is not necessarily the case that a material in a particular place is in a state that is steady and nearly homogeneous enough to allow it to have a well-defined hotness or temperature. Hotness may be represented abstractly as a one-dimensional manifold. Every valid temperature scale has its own one-to-one map into the hotness manifold.

When two systems in thermal contact are at the same temperature no heat transfers between them. When a temperature difference does exist heat flows spontaneously from the warmer system to the colder system until they are in thermal equilibrium. Heat transfer occurs by conduction or by thermal radiation.

Experimental physicists, for example Galileo and Newton, found that there are indefinitely many empirical temperature scales. Nevertheless, the zeroth law of thermodynamics says that they all measure the same quality.

Temperature for Bodies in Thermodynamic Equilibrium

For experimental physics, hotness means that, when comparing any two given bodies in their respective separate thermodynamic equilibria, any two suitably given empirical thermometers with numerical scale readings will agree as to which is the hotter of the two given bodies, or that they have the same temperature. This does not require the two thermometers to have a linear relation between their numerical scale readings, but it does require that the relation between their numerical readings shall be strictly monotonic. A definite sense of greater hotness can be had, independently of calorimetry, of thermodynamics, and of properties of particular materials, from Wien's displacement law of thermal radiation: the temperature of a bath of thermal radiation is proportional, by a universal constant, to the frequency of the maximum of its frequency spectrum; this frequency is always positive, but can have values that tend to zero. Thermal radiation is initially defined for a cavity in thermodynamic equilibrium. These physical facts justify a mathematical statement that hotness exists on an ordered one-dimensional manifold. This is a fundamental character of temperature and thermometers for bodies in their own thermodynamic equilibrium.

Except for a system undergoing a first-order phase change such as the melting of ice, as a closed system receives heat, without change in its volume and without change in external force fields acting on it, its temperature rises. For a system undergoing such a phase change so slowly that departure from thermodynamic equilibrium can be neglected, its temperature remains constant as the system is supplied with latent heat. Conversely, a loss of heat from a closed system, without phase change, without change of volume, and without change in external force fields acting on it, decreases its temperature.

Temperature for Bodies in a Steady State But not in Thermodynamic equilibrium

While for bodies in their own thermodynamic equilibrium states, the notion of temperature requires that all empirical thermometers must agree as to which of two bodies is the hotter or that they are at the same temperature, this requirement is not safe for bodies that are in steady states though not in thermodynamic equilibrium. It can then well be that different empirical thermometers disagree about which is the hotter, and if this is so, then at least one of the bodies does not have a well defined absolute thermodynamic temperature. Nevertheless, any one given body and any one suitable empirical thermometer can still support notions of empirical, non-absolute, hotness and temperature, for a suitable range of processes. This is a matter for study in non-equilibrium thermodynamics.

Temperature for Bodies not in a Steady State

When a body is not in a steady state, then the notion of temperature becomes even less safe than for a body in a steady state not in thermodynamic equilibrium. This is also a matter for study in non-equilibrium thermodynamics.

Thermodynamic Equilibrium Axiomatics

For axiomatic treatment of thermodynamic equilibrium, since the 1930s, it has become customary to refer to a zeroth law of thermodynamics. The customarily stated minimalist version of such a law postulates only that all bodies, which when thermally connected would be in thermal equilibrium, should be said to have the same temperature by definition, but by itself does not establish temperature as a quantity expressed as a real number on a scale. A more physically informative version of such a law views empirical temperature as a chart on a hotness manifold. While the zeroth law permits the definitions of many different empirical scales of temperature, the second law of thermodynamics selects the definition of a single preferred, absolute temperature, unique up to an arbitrary scale factor, whence called the thermodynamic temperature. If internal energy is considered as a function of the volume and entropy of a homogeneous system in thermodynamic equilibrium, thermodynamic absolute temperature appears as the partial derivative of internal energy with respect the entropy at constant volume. Its natural, intrinsic origin or null point is absolute zero at which the entropy of any system is at a minimum. Although this is the lowest absolute temperature described by the model, the third law of thermodynamics postulates that absolute zero cannot be attained by any physical system.

Heat Capacity

When an energy transfer to or from a body is only as heat, state of the body changes. Depending on the surroundings and the walls separating them from the body, various changes are possible in the body. They include chemical reactions, increase of pressure, increase of temperature, and phase change. For each kind of change under specified conditions, the heat capacity is the ratio of the quantity of heat transferred to the magnitude of the change. For example, if the change is an increase in temperature at constant volume, with no phase change and no chemical change, then the temperature of the body rises and its pressure increases. The quantity of heat transferred, ΔQ, divided by the observed temperature change, ΔT, is the body's heat capacity at constant volume, C_V.

$$C_V = \frac{\Delta Q}{\Delta T}$$

If heat capacity is measured for a well defined amount of substance, the specific heat is the measure of the heat required to increase the temperature of such a unit quantity by one unit of temperature. For example, to raise the temperature of water by one kelvin (equal to one degree Celsius) requires 4186 joules per kilogram (J/kg).

Temperature Measurement

A typical Celsius thermometer measures a winter day temperature of −17°C

Temperature measurement using modern scientific thermometers and temperature scales goes back at least as far as the early 18th century, when Gabriel Fahrenheit adapted a thermometer (switching to mercury) and a scale both developed by Ole Christensen Rømer. Fahrenheit's scale is still in use in the United States for non-scientific applications.

Temperature is measured with thermometers that may be calibrated to a variety of temperature scales. In most of the world (except for Belize, Myanmar, Liberia and the United States), the Celsius scale is used for most temperature measuring purposes. Most scientists measure temperature using the Celsius scale and thermodynamic temperature using the Kelvin scale, which is the Celsius scale offset so that its null point is 0K = −273.15°C, or absolute zero. Many engineering fields in the U.S., notably high-tech and US federal specifications (civil and military), also use the Kelvin and Celsius scales. Other engineering fields in the U.S. also rely upon the Rankine scale (a shifted Fahrenheit scale) when working in thermodynamic-related disciplines such as combustion.

Units

The basic unit of temperature in the International System of Units (SI) is the kelvin. It has the symbol K.

For everyday applications, it is often convenient to use the Celsius scale, in which 0°C corresponds very closely to the freezing point of water and 100°C is its boiling point at sea level. Because liquid droplets commonly exist in clouds at sub-zero temperatures, 0°C is better defined as the melting point of ice. In this scale a temperature difference of 1 degree Celsius is the same as a 1kelvin increment, but the scale is offset by the temperature at which ice melts (273.15 K).

By international agreement the Kelvin and Celsius scales are defined by two fixing points: absolute zero and the triple point of Vienna Standard Mean Ocean Water, which is water specially prepared with a specified blend of hydrogen and oxygen isotopes. Absolute zero is defined as precisely 0K and −273.15°C. It is the temperature at which all classical translational motion of the particles comprising matter ceases and they are at complete rest in the classical model. Quantum-mechanically, however, zero-point motion remains and has an associated energy, the zero-point energy. Matter is in its ground state, and contains no thermal energy. The triple point of water is defined as 273.16K and 0.01°C. This definition serves the following purposes: it fixes the magnitude of the kelvin as being precisely 1 part in 273.16 parts of the difference between absolute zero and the triple point of water; it establishes that one kelvin has precisely the same magnitude as one degree on the Celsius scale; and it establishes the difference between the null points of these scales as being 273.15K (0K = −273.15°C and 273.16K = 0.01°C).

In the United States, the Fahrenheit scale is widely used. On this scale the freezing point of water corresponds to 32 °F and the boiling point to 212 °F. The Rankine scale, still used in fields of chemical engineering in the U.S., is an absolute scale based on the Fahrenheit increment.

Conversion

The following table shows the temperature conversion formulas for conversions to and from the Celsius scale.

	from Celsius	to Celsius
Fahrenheit	[°F] = [°C] × $\frac{9}{5}$ + 32	[°C] = ([°F] − 32) × $\frac{5}{9}$
Kelvin	[K] = [°C] + 273.15	[°C] = [K] − 273.15
Rankine	[°R] = ([°C] + 273.15) × $\frac{9}{5}$	[°C] = ([°R] − 491.67) × $\frac{5}{9}$
Delisle	[°De] = (100 − [°C]) × $\frac{3}{2}$	[°C] = 100 − [°De] × $\frac{2}{3}$
Newton	[°N] = [°C] × $\frac{33}{100}$	[°C] = [°N] × $\frac{100}{33}$
Réaumur	[°Ré] = [°C] × $\frac{4}{5}$	[°C] = [°Ré] × $\frac{5}{4}$
Rømer	[°Rø] = [°C] × $\frac{21}{40}$ + 7.5	[°C] = ([°Rø] − 7.5) × $\frac{40}{21}$

Plasma Physics

The field of plasma physics deals with phenomena of electromagnetic nature that involve very high temperatures. It is customary to express temperature as energy in units of electronvolts (eV) or kiloelectronvolts (keV). The energy, which has a different dimension from temperature, is then calculated as the product of the Boltzmann constant and temperature, $E = k_B T$. Then, 1 eV corresponds to 11605K. In the study of QCD matter one routinely encounters temperatures of the order of a few hundred MeV, equivalent to about 10^{12}K.

Theoretical Foundation

Historically, there are several scientific approaches to the explanation of temperature: the classical thermodynamic description based on macroscopic empirical variables that can be measured in a laboratory; the kinetic theory of gases which relates the macroscopic description to the probability distribution of the energy of motion of gas particles; and a microscopic explanation based on statistical physics and quantum mechanics. In addition, rigorous and purely mathematical treatments have provided an axiomatic approach to classical thermodynamics and temperature. Statistical physics provides a deeper understanding by describing the atomic behavior of matter, and derives macroscopic properties from statistical averages of microscopic states, including both classical and quantum states. In the fundamental physical description, using natural units, temperature may be measured directly in units of energy. However, in the practical systems of measurement for science, technology, and commerce, such as the modern metric system of units, the macroscopic and the microscopic descriptions are interrelated by the Boltzmann constant, a proportionality factor that scales temperature to the microscopic mean kinetic energy.

The microscopic description in statistical mechanics is based on a model that analyzes a system into its fundamental particles of matter or into a set of classical or quantum-mechanical oscillators and considers the system as a statistical ensemble of microstates. As a collection of classical material particles, temperature is a measure of the mean energy of motion, called kinetic energy, of the particles, whether in solids, liquids, gases, or plasmas. The kinetic energy, a concept of classical mechanics, is half the mass of a particle times its speed squared. In this mechanical interpretation of thermal motion, the kinetic energies of material particles may reside in the velocity of the particles of their translational or vibrational motion or in the inertia of their rotational modes. In monatomic perfect gases and, approximately, in most gases, temperature is a measure of the mean particle kinetic energy. It also determines the probability distribution function of the energy. In condensed matter, and particularly in solids, this purely mechanical description is often less useful and the oscillator model provides a better description to account for quantum mechanical phenomena. Temperature determines the statistical occupation of the microstates of the ensemble. The microscopic definition of temperature is only meaningful in the thermodynamic limit, meaning for large ensembles of states or particles, to fulfill the requirements of the statistical model.

In the context of thermodynamics, the kinetic energy is also referred to as thermal energy. The thermal energy may be partitioned into independent components attributed to the degrees of freedom of the particles or to the modes of oscillators in a thermodynamic system. In general, the number of these degrees of freedom that are available for the equipartitioning of energy depend on the temperature, i.e. the energy region of the interactions under consideration. For solids, the thermal energy is associated primarily with the vibrations of its atoms or molecules about their equilibrium position. In an ideal monatomic gas, the kinetic energy is found exclusively in the purely translational motions of the particles. In other systems, vibrational and rotational motions also contribute degrees of freedom.

Kinetic Theory of Gases

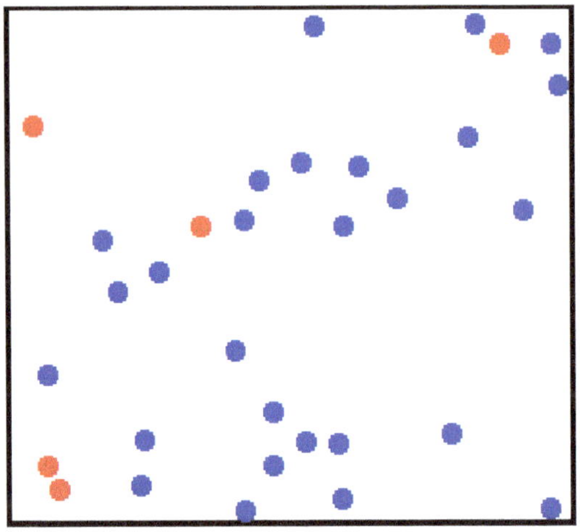

A theoretical understanding of temperature in an ideal gas can be obtained from the Kinetic theory.

Maxwell and Boltzmann developed a kinetic theory that yields a fundamental understanding of temperature in gases. This theory also explains the ideal gas law and the observed heat capacity of monatomic (or 'noble') gases.

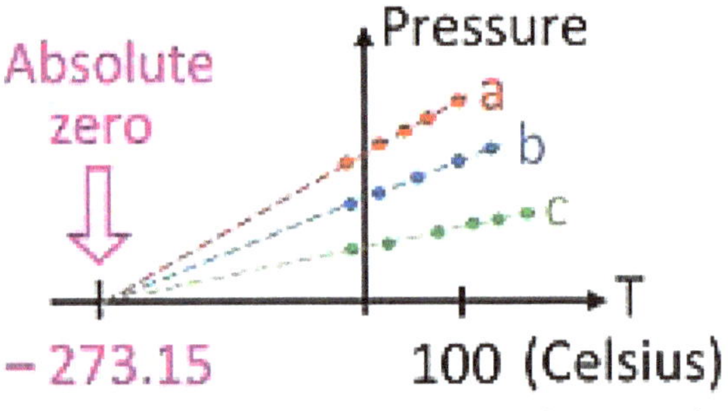

Plots of pressure vs temperature for three different gas samples extrapolated to absolute zero.

The ideal gas law is based on observed empirical relationships between pressure (p), volume (V), and temperature (T), and was recognized long before the kinetic theory of gases was developed. The ideal gas law states:

$$pV = nRT$$

where n is the number of moles of gas and $R = 8.3144598(48)\ \mathrm{J \cdot mol^{-1} \cdot K^{-1}}$ is the gas constant.

This relationship gives us our first hint that there is an absolute zero on the temperature scale, because it only holds if the temperature is measured on an absolute scale such as Kelvins. The ideal gas law allows one to measure temperature on this absolute scale using the gas thermometer. The temperature in kelvins can be defined as the pressure in pascals of one mole of gas in a container of one cubic meter, divided by the gas constant.

Although it is not a particularly convenient device, the gas thermometer provides an essential theoretical basis by which all thermometers can be calibrated. As a practical matter it is not possible to use a gas thermometer to measure absolute zero temperature since the gases tend to condense into a liquid long before the temperature reaches zero. It is possible, however, to extrapolate to absolute zero by using the ideal gas law, as shown in the figure.

The kinetic theory assumes that pressure is caused by the force associated with individual atoms striking the walls, and that all energy is translational kinetic energy. Using a sophisticated symmetry argument, Boltzmann deduced what is now called the Maxwell–Boltzmann probability distribution function for the velocity of particles in an ideal gas. From that probability distribution function, the average kinetic energy, E_k (per particle), of a monatomic ideal gas is:

$$E_{\mathrm{k}} = \frac{1}{2} m v_{\mathrm{rms}}^2 = \frac{3}{2} kT,$$

where the Boltzmann constant, k, is the ideal gas constant divided by the Avogadro number, and v_{rms} is the root-mean-square speed. Thus the ideal gas law states that internal energy is directly proportional to temperature. This direct proportionality between temperature and internal energy is a special case of the equipartition theorem, and holds only in the classical limit of an ideal gas. It does not hold for most substances, although it is true that temperature is a monotonic (non-decreasing) function of internal energy.

Zeroth Law of Thermodynamics

When two otherwise isolated bodies are connected together by a rigid physical path impermeable to matter, there is spontaneous transfer of energy as heat from the hotter to the colder of them. Eventually they reach a state of mutual thermal equilibrium, in which heat transfer has ceased, and the bodies' respective state variables have settled to become unchanging.

One statement of the zeroth law of thermodynamics is that if two systems are each in thermal equilibrium with a third system, then they are also in thermal equilibrium with each other.

This statement helps to define temperature but it does not, by itself, complete the definition. An empirical temperature is a numerical scale for the hotness of a thermodynamic system. Such hotness may be defined as existing on a one-dimensional manifold, stretching between hot and cold. Sometimes the zeroth law is stated to include the existence of a unique universal hotness manifold, and of numerical scales on it, so as to provide a complete definition of empirical temperature. To be suitable for empirical thermometry, a material must have a monotonic relation between hotness and some easily measured state variable, such as pressure or volume, when all other relevant coordinates are fixed. An exceptionally suitable system is the ideal gas, which can provide a temperature scale that matches the absolute Kelvin scale. The Kelvin scale is defined on the basis of the second law of thermodynamics.

Second law of Thermodynamics

It is also possible to define temperature in terms of the second law of thermodynamics which deals with entropy. The second law states that any process will result in either no change or a net increase in the entropy of the universe. This can be understood in terms of probability.

For example, in a series of coin tosses, a perfectly ordered system would be one in which either every toss comes up heads or every toss comes up tails. This means that for a perfectly ordered set of coin tosses, there is only one set of toss outcomes possible: the set in which 100% of tosses come up the same. On the other hand, there are multiple combinations that can result in disordered or mixed systems, where some fraction are heads and the rest tails. A disordered system can be 90% heads and 10% tails, or it could be 98% heads and 2% tails, et cetera. As the number of coin tosses increases, the number of possible combinations corresponding to imperfectly ordered systems increases. For a very large number of coin tosses, the combinations to ~50% heads and ~50% tails dominates and obtaining an outcome significantly different from 50/50 becomes extremely unlikely. Thus the system naturally progresses to a state of maximum disorder or entropy.

It has been previously stated that temperature governs the transfer of heat between two systems and it was just shown that the universe tends to progress so as to maximize entropy, which is expected of any natural system. Thus, it is expected that there is some relationship between temperature and entropy. To find this relationship, the relationship between heat, work and temperature is first considered. A heat engine is a device for converting thermal energy into mechanical energy, resulting in the performance of work, and analysis of the Carnot heat engine provides the necessary relationships. The work from a heat engine corresponds to the difference between the heat put into the system at the high temperature, q_H and the heat ejected at the low temperature, q_C. The efficiency is the work divided by the heat put into the system or:

$$\text{efficiency} = \frac{w_{cy}}{q_H} = \frac{q_H - q_C}{q_H} = 1 - \frac{q_C}{q_H} \qquad (4)$$

where w_{cy} is the work done per cycle. The efficiency depends only on q_C/q_H. Because q_C and q_H correspond to heat transfer at the temperatures T_C and T_H, respectively, q_C/q_H should be some function of these temperatures:

$$\frac{q_C}{q_H} = f(T_H, T_C) \qquad (5)$$

Carnot's theorem states that all reversible engines operating between the same heat reservoirs are equally efficient. Thus, a heat engine operating between T_1 and T_3 must have the same efficiency as one consisting of two cycles, one between T_1 and T_2, and the second between T_2 and T_3. This can only be the case if:

$$q_{13} = \frac{q_1 q_2}{q_2 q_3}$$

which implies:

$$q_{13} = f(T_1, T_3) = f(T_1, T_2) f(T_2, T_3)$$

Since the first function is independent of T_2, this temperature must cancel on the right side, meaning $f(T_1,T_3)$ is of the form $g(T_1)/g(T_3)$ (i.e. $f(T_1,T_3) = f(T_1,T_2)f(T_2,T_3) = g(T_1)/g(T_2)\cdot g(T_2)/g(T_3) = g(T_1)/g(T_3)$), where g is a function of a single temperature. A temperature scale can now be chosen with the property that:

$$\frac{q_C}{q_H} = \frac{T_C}{T_H} \qquad (6)$$

Substituting Equation 4 back into Equation 2 gives a relationship for the efficiency in terms of temperature:

$$\text{efficiency} = 1 - \frac{q_C}{q_H} = 1 - \frac{T_C}{T_H} \qquad (7)$$

For $T_C = 0$ K the efficiency is 100% and that efficiency becomes greater than 100% below 0 K. Since an efficiency greater than 100% violates the first law of thermodynamics, this implies that 0 K is the minimum possible temperature. In fact the lowest temperature ever obtained in a macroscopic system was 20 nK, which was achieved in 1995 at NIST. Subtracting the right hand side of Equation 5 from the middle portion and rearranging gives:

$$\frac{q_H}{T_H} - \frac{q_C}{T_C} = 0$$

where the negative sign indicates heat ejected from the system. This relationship suggests the existence of a state function, S, defined by:

$$dS = \frac{dq_{\text{rev}}}{T} \qquad (8)$$

where the subscript indicates a reversible process. The change of this state function around any cycle is zero, as is necessary for any state function. This function corresponds to the entropy of the system, which was described previously. Rearranging Equation 6 gives a new definition for temperature in terms of entropy and heat:

$$T = \frac{dq_{\text{rev}}}{dS} \qquad (9)$$

For a system, where entropy $S(E)$ is a function of its energy E, the temperature T is given by:

$$T^{-1} = \frac{d}{dE}S(E) \qquad (10)$$

i.e. the reciprocal of the temperature is the rate of increase of entropy with respect to energy.

Definition from Statistical Mechanics

Statistical mechanics defines temperature based on a system's fundamental degrees of freedom. Eq.(10) is the defining relation of temperature. Eq. (9) can be derived from the principles underlying the fundamental thermodynamic relation.

Generalized Temperature from Single-particle Statistics

It is possible to extend the definition of temperature even to systems of few particles, like in a quantum dot. The generalized temperature is obtained by considering time ensembles instead of configuration-space ensembles given in statistical mechanics in the case of thermal and particle exchange between a small system of fermions (N even less than 10) with a single/double-occupancy system. The finite quantum grand canonical ensemble, obtained under the hypothesis of ergodicity and orthodicity, allows to express the generalized temperature from the ratio of the average time of occupation τ_1 and τ_2 of the single/double-occupancy system:

$$T = k^{-1} \ln 2 \frac{\tau_2}{\tau_1} \left(E - E_{\mathrm{F}} \left(1 + \frac{3}{2N} \right) \right),$$

where E_{F} is the Fermi energy. This generalized temperature tends to the ordinary temperature when N goes to infinity.

Negative Temperature

On the empirical temperature scales, which are not referenced to absolute zero, a negative temperature is one below the zero-point of the scale used. For example, dry ice has a sublimation temperature of −78.5°C which is equivalent to −109.3°F. On the absolute Kelvin scale, however, this temperature is 194.6 K. On the absolute scale of thermodynamic temperature no material can have a temperature smaller than or equal to 0 K, both of which are forbidden by the third law of thermodynamics.

Temperature is basically defined for a body in its own state of internal thermodynamic equilibrium, and in this definition, on an absolute scale, it is always positive. In an apparent contradiction of this reliable and valid rule, a so-called negative absolute "temperature" may be approximately defined for a component of a body that is not in its own state of internal thermodynamic equilibrium: a component may have a negative approximate "temperature" while the rest of the components of the body have positive approximate temperatures. Such a non-equilibrium situation is either transient in time or is maintained by external factors that drive a flow of energy through the body of interest. An example of such a component is a spin system within a body, as follows.

In the quantum mechanical description of electron and nuclear spin systems that have a limited number of possible states, and therefore a discrete upper limit of energy they can attain, it is possible to obtain a negative temperature, which is numerically indeed less than absolute zero. However, this is not the macroscopic temperature of the material, but instead the temperature of only very specific degrees of freedom, that are isolated from others and do not exchange energy by virtue of the equipartition theorem.

A negative temperature is experimentally achieved with suitable radio frequency techniques that cause a population inversion of spin states from the ground state. As the energy in the system increases upon population of the upper states, the entropy increases as well, as the system becomes less ordered, but attains a maximum value when the spins are evenly distributed among ground and excited states, after which it begins to decrease, once again achieving a state of higher order as the upper states begin to fill exclusively. At the point of maximum entropy, the temperature function shows the behavior of a singularity, because the slope of the entropy function decreases to zero at first and then turns negative. Since temperature is the inverse of the derivative of the entropy, the temperature formally goes to infinity at this point, and switches to negative infinity as the slope turns negative. At energies higher than this point, the spin degree of freedom therefore exhibits formally a negative thermodynamic temperature. As the energy increases further by continued population of the excited state, the negative temperature approaches zero asymptotically. As the energy of the system increases in the population inversion, a system with a negative temperature is not colder than absolute zero, but rather it has a higher energy than at positive temperature, and may be said to be in fact hotter at negative temperatures. When brought into contact with a system at a positive temperature, energy will be transferred from the negative temperature regime to the positive temperature region.

Examples of Temperature

	Temperature		Peak emittance wavelength of black-body radiation
	Kelvin	Celsius	
Absolute zero (precisely by definition)	0 K	−273.15 °C	cannot be defined
Coldest temperature achieved	100 pK	−273.149999999900 °C	29,000 km
Coldest Bose–Einstein condensate	450 pK	−273.14999999955 °C	6,400 km
One millikelvin (precisely by definition)	0.001 K	−273.149 °C	2.89777 m (radio, FM band)
Cosmic microwave background (2013 measurement)	2.7260 K	−270.424 °C	0.00106301 m (millimeter-wavelength microwave)
Water triple point (precisely by definition)	273.16 K	0.01 °C	10,608.3 nm (long-wavelength IR)
Water boiling point[A]	373.1339 K	99.9839 °C	7,766.03 nm (mid-wavelength IR)
Incandescent lamp[B]	2500 K	≈2,200 °C	1,160 nm (near infrared)[C]
Sun's visible surface[D]	5,778 K	5,505 °C	501.5 nm (green-blue light)
Lightning bolt channel[E]	28 kK	28,000 °C	100 nm (far ultraviolet light)
Sun's core[E]	16 MK	16 million °C	0.18 nm (X-rays)
Thermonuclear weapon (peak temperature)[E]	350 MK	350 million °C	8.3×10^{-3} nm (gamma rays)
Sandia National Labs' Z machine[E]	2 GK	2 billion °C	1.4×10^{-3} nm (gamma rays)[F]
Core of a high-mass star on its last day[E]	3 GK	3 billion °C	1×10^{-3} nm (gamma rays)
Merging binary neutron star system[E]	350 GK	350 billion °C	8×10^{-6} nm (gamma rays)
Relativistic Heavy Ion Collider[E]	1 TK	1 trillion °C	3×10^{-6} nm (gamma rays)

CERN's proton vs nucleus collisions[E]	10 TK	10 trillion °C	3×10^{-7} nm (gamma rays)
Universe 5.391×10^{-44} s after the Big Bang[E]	1.417×10^{32} K	1.417×10^{32} °C	1.616×10^{-27} nm (Planck length)

- For Vienna Standard Mean Ocean Water at one standard atmosphere (101.325 kPa) when calibrated strictly per the two-point definition of thermodynamic temperature.
- The 2500 K value is approximate. The 273.15 K difference between K and °C is rounded to 300 K to avoid false precision in the Celsius value.
- For a true black-body (which tungsten filaments are not). Tungsten filament emissivity is greater at shorter wavelengths, which makes them appear whiter.
- Effective photosphere temperature. The 273.15 K difference between K and °C is rounded to 273 K to avoid false precision in the Celsius value.
- The 273.15 K difference between K and °C is within the precision of these values.
- For a true black-body (which the plasma was not). The Z machine's dominant emission originated from 40 MK electrons (soft x-ray emissions) within the plasma.

Temperature is generally considered as indicative of quantity of heat. It is equivalent to potential in electricity and level in hydrostatics. On the basis of kinetic theory, the temperature, may be defined as

$$\frac{1}{2}mV_{av}^2=kT$$

where, m is mass of molecule.

V_{av} is the average velocity and

k is the Boltzmans constant

The above definition is applicable only for systems which obey Maxwell-Boltzman distribution. Some authors define it as a condition of a body by virtue of which heat is transferred to or from other bodies.

Temperature is measured by the observation of certain of the properties of matter which are influenced by the degree of heat. The most used are changes in

1. Physical state
2. Chemical state
3. Dimensions
4. Electrical properties
5. Radiation properties

The instruments to measure temperature have been classified according to the nature of change produced in the testing body by change in temperature. Based on the above consideration there are four categories of thermometers .They are:

1. Expansion Thermometers
2. Change of State Thermometers
3. Electrical Thermometers
4. Radiation and Optical pyrometers

Expansion Thermometers or Liquid in Glass Thermometers (LIG)

A liquid-in-glass thermometer is widely used due to its accuracy for the temperature range -200 to 600°C. Compared to other thermometers, it is simple for usage. It has been used in medicine, metrology and industry. In the LIG thermometer the thermally sensitive element is a liquid contained in a graduated glass envelope. The liquid used in practical thermometers are mercury or alcohol. The principle used to measure temperature is that of the apparent thermal expansion of the liquid.

Boiling point of Mercury is 357 °C. Liquid in bulb Thermometers making use of Mercury has a range of -39° to 35°C which are freezing and melting points of Mercury. If alcohol is used the lower temperature can be upto -62 °C

Bimetallic Thermometers

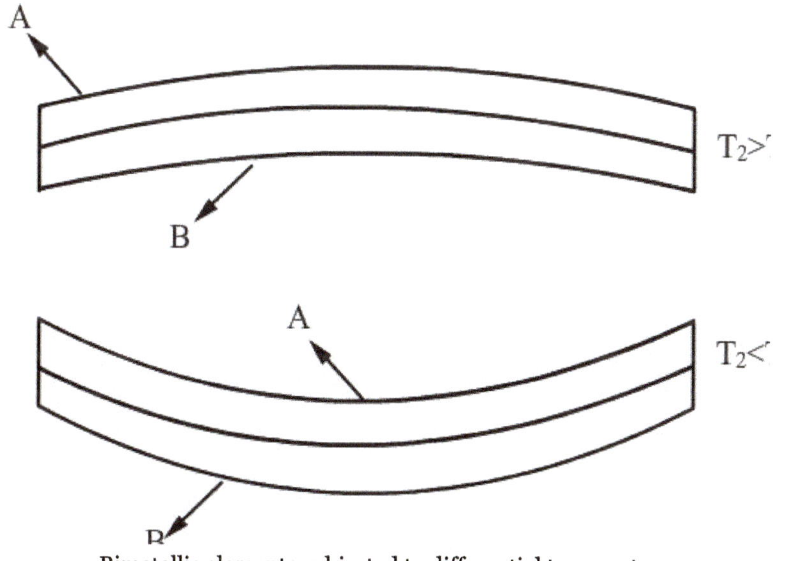

Bimetallic elements subjected to differential temperature

Bimetallic elements are used for temperature measurement and more widely for sensing and control purposes Invar rod (coefficient of expansion =2.7x 10^{-6} cm/°C) and Brass (coefficient of expansion = 34.2x10^{-6} cm/°C because of their large difference in the values of coefficient of expansion are used as a very useful combination.

Change of State Thermometers: Vapor Pressure Thermometer

Consider a container having a certain quantity of liquid. The molecules of liquid are in a state of random motion moving in all directions. When vertical component of kinetic energy is greater than the force of attraction at liquid surface, it escapes from the surface. On the other hand, the molecules which constitute vapour are also moving at random. The process of evaporation and condensation go on simultaneously. When the rates of evaporation and condensation are equal, the vapour becomes saturated.

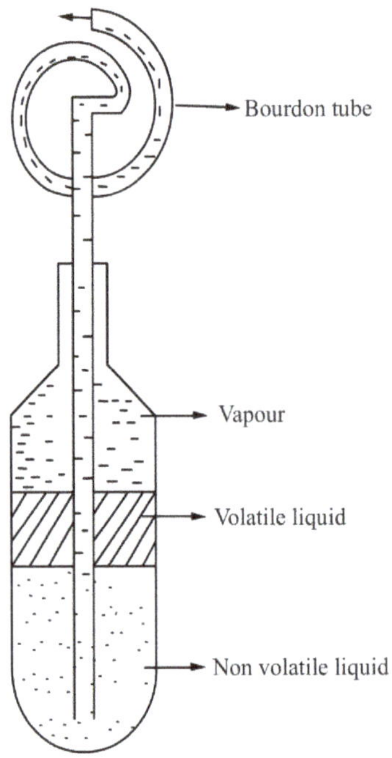

Vapor pressure thermometer

Saturated vapor pressure depends only on the temperature and properties of the liquid and is independent of the size of the container. The vapor pressure thermometer consists of a (i) bulb containing a fill liquid (ii) a capillary tube (iii) Bourdon Tube.

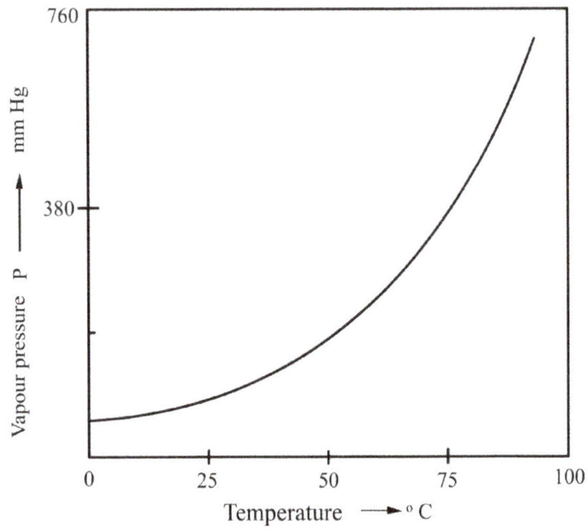

Saturated vapor pressure of water

The relationship between vapor pressure and temperature is non-linear. As an example, the relation between the saturated vapor pressure of water is shown in the figure above. The range of usage of the vapor pressure thermometer is dependent on the saturated vapor pressure of the liquid. Some of the commonly used fill liquids and the range of temperature in degree C (given in brackets) are Methyl chloride (0-50), Sulphur dioxide (30-120),Water (120-220), Butane (20-80) and Toluene (150-250). The bulb of the thermometer is exposed to the temperature field to be measured.

Electrical Resistance Thermometry

In this category there are two types (i) those using conductor sensors (ii) those using semiconductors (thermistors)

Conductor Sensors

Resistance of pure metallic conductors increases with temperature in a reproducible manner. Some of the metals and the range of temperature measurement using them are the following

Platinum - 190°C to 630.5°C

Copper - 50°C to 250°C

Nickel - 200°C to 350°C

Platinum is the preferred metal because of the following reasons:

a) Platinum is stable

b) Can be drawn to fine wires

c) Available in high purity

The resistance temperature relationship is not linear in the complete temperature range. The resistance at any particular temperature may be written as

$$R = R_0\left[1+\alpha_1 T+\alpha_2 T^2+.............\alpha_n T^n\right]$$

where R_0 is the resistance at temperature T=0. The number of terms necessary depends on the material, the accuracy required and the temperature range covered. Platinum, nickel and copper require respectively two, three and three of the α constants for accurate representation. Platinum, for instance, is linear within +/- 0.2% from 255K to 366K, +/- 0.4% in the range 90K to 183K and also in the range 200K to 422K.

If assumed linear,

$R_t = R_0 (1+\alpha t)$

$\alpha \rightarrow$ (Temperature coefficient of resistance)

(t is any temperature)

α is obtained by measuring resistances R_{100} and R_0 at steam and ice.

Thermistor

A thermistor is a type of resistor whose resistance is dependent on temperature, more so than in standard resistors. The word is a portmanteau of *thermal* and *resistor*. Thermistors are widely used as inrush current limiter, temperature sensors (Negative Temperature Coefficient or NTC type typically), self-resetting overcurrent protectors, and self-regulating heating elements (Positive Temperature Coefficient or PTC type typically).

Thermistors are of two opposite fundamental types:

- With NTC, resistance *decreases* as temperature rises to protect against inrush *overvoltage* conditions. Commonly installed in parallel as a current sink.
- With PTC, resistance *increases* as temperature rises to protect against *overcurrent* conditions. Commonly installed in series as a resettable fuse.

Thermistors differ from resistance temperature detectors (RTDs) in that the material used in a thermistor is generally a ceramic or polymer, while RTDs use pure metals. The temperature response is also different; RTDs are useful over larger temperature ranges, while thermistors typically achieve a greater precision within a limited temperature range, typically −90 °C to 130 °C.

Basic Operation

Assuming, as a first-order approximation, that the relationship between resistance and temperature is linear, then:

$$\Delta R = k\Delta T$$

where

ΔR, change in resistance

ΔT, change in temperature

k, first-order temperature coefficient of resistance

Thermistors can be classified into two types, depending on the classification of k. If k is positive, the resistance increases with increasing temperature, and the device is called a positive temperature coefficient (PTC) thermistor, or posistor. If k is negative, the resistance decreases with increasing temperature, and the device is called a negative temperature coefficient (NTC) thermistor. Resistors that are not thermistors are designed to have a k as close to 0 as possible, so that their resistance remains nearly constant over a wide temperature range.

Instead of the temperature coefficient *k*, sometimes the *temperature coefficient of resistance* α_T (alpha sub T) is used. It is defined as

$$\alpha_T = \frac{1}{R(T)}\frac{dR}{dT}.$$

This α_T coefficient should not be confused with the a parameter below.

Steinhart–Hart Equation

In practice, the linear approximation (above) works only over a small temperature range. For accurate temperature measurements, the resistance/temperature curve of the device must be described in more detail. The Steinhart–Hart equation is a widely used third-order approximation:

$$\frac{1}{T} = a + b\ln(R) + c(\ln(R))^3$$

where a, b and c are called the Steinhart–Hart parameters, and must be specified for each device. T is the absolute temperature and R is the resistance. To give resistance as a function of temperature, the above can be rearranged into:

$$R = \exp\left[\left(x - \frac{1}{2}y\right)^{\frac{1}{3}} - \left(x + \frac{1}{2}y\right)^{\frac{1}{3}}\right]$$

where

$$y = \frac{1}{c}\left(a - \frac{1}{T}\right) \qquad x = \sqrt{\left(\frac{b}{3c}\right)^3 + \left(\frac{y}{2}\right)^2}$$

The error in the Steinhart–Hart equation is generally less than 0.02 °C in the measurement of temperature over a 200 °C range. As an example, typical values for a thermistor with a resistance of 3 kΩ at room temperature (25 °C = 298.15 K) are:

$$a = 1.40\times10^{-3} \qquad b = 2.37\times10^{-4} \qquad c = 9.90\times10^{-8}$$

B or *β* Parameter Equation

NTC thermistors can also be characterised with the B (or β) parameter equation, which is essentially the Steinhart–Hart equation with $a = (1/T_0) - (1/B)\ln(R_0)$, $b = 1/B$ and $c = 0$,

$$\frac{1}{T} = \frac{1}{T_0} + \frac{1}{B}\ln\left(\frac{R}{R_0}\right),$$

where the temperatures are in kelvins, and R_0 is the resistance at temperature T_0 (25 °C = 298.15 K). Solving for R yields:

$$R = R_0 e^{-B\left(\frac{1}{T_0} - \frac{1}{T}\right)}$$

or, alternatively,

$$R = r_\infty e^{B/T}$$

where $r_\infty = R_0 e^{-B/T_0}$.

This can be solved for the temperature:

$$T = \frac{B}{\ln(R / r_\infty)}$$

The B-parameter equation can also be written as $\ln R = B / T + \ln r_\infty$. This can be used to convert the function of resistance vs. temperature of a thermistor into a linear function of $\ln R$ vs. $1/T$. The average slope of this function will then yield an estimate of the value of the B parameter.

Conduction Model

NTC (Negative Temperature Coefficient)

Many NTC thermistors are made from a pressed disc, rod, plate, bead or cast chip of semiconducting material such as sintered metal oxides. They work because raising the temperature of a semiconductor increases the number of active charge carriers - it promotes them into the *conduction band.* The more charge carriers that are available, the more current a material can conduct. In certain materials like ferric oxide (Fe_2O_3) with titanium (Ti) doping an *n-type* semiconductor is formed and the charge carriers are electrons. In materials such as nickel oxide (NiO) with lithium (Li) doping a *p-type* semiconductor is created where holes are the charge carriers.

This is described in the formula:

$$I = n \cdot A \cdot v \cdot e$$

I = electric current (amperes)

n = density of charge carriers (count/m^3)

A = cross-sectional area of the material (m^2)

v = drift velocity of electrons (m/s)

e = charge of an electron ($e = 1.602 \times 10^{-19}$ coulomb)

Over large changes in temperature, calibration is necessary. Over small changes in temperature, if the right semiconductor is used, the resistance of the material is linearly proportional to the temperature. There are many different semiconducting thermistors with a range from about 0.01 kelvin to 2,000 kelvins (−273.14 °C to 1,700 °C).

PTC (Positive Temperature Coefficient)

Most PTC thermistors are made from doped polycrystalline ceramic (containing barium titanate ($BaTiO_3$) and other compounds) which have the property that their resistance rises suddenly at a certain critical temperature. Barium titanate is ferroelectric and its dielectric constant varies with temperature. Below the Curie point temperature, the high dielectric constant prevents the formation of potential barriers between the crystal grains, leading to a low resistance. In this region the device has a small negative temperature coefficient. At the Curie point temperature, the dielectric constant drops sufficiently to allow the formation of potential barriers at the grain boundaries, and the resistance increases sharply with temperature. At even higher temperatures, the material reverts to NTC behaviour.

Another type of thermistor is a silistor, a thermally sensitive silicon resistor. Silistors employ silicon as the semiconductive component material. Unlike ceramic PTC thermistors, silistors have an almost linear resistance-temperature characteristic.

Barium titanate thermistors can be used as self-controlled heaters; for a given voltage, the ceramic will heat to a certain temperature, but the power used will depend on the heat loss from the ceramic.

The dynamics of PTC thermistors being powered also is extremely useful. When first connected to a voltage source, a large current corresponding to the low, cold, resistance flows, but as the thermistor self-heats, the current is reduced until a limiting current (and corresponding peak device temperature) is reached. The current-limiting effect can replace fuses. They are also used in the degaussing circuits of many CRT monitors and televisions where the degaussing coil only has to be connected in series with an appropriately chosen thermistor; a particular advantage is that the current decrease is smooth, producing optimum degausing effect. Improved degaussing circuits have auxiliary heating elements to heat the thermistor further (and reduce the final current) or timed relays to disconnect the degaussing circuit entirely after it has operated.

Another type of PTC thermistor is the polymer PTC, which is sold under brand names such as "Polyswitch" "Semifuse", and "Multifuse". This consists of plastic with carbon grains embedded in it. When the plastic is cool, the carbon grains are all in contact with each other, forming a conductive path through the device. When the plastic heats up, it expands, forcing the carbon grains apart, and causing the resistance of the device to rise, which then causes increased heating and rapid resistance increase. Like the $BaTiO_3$ thermistor, this device has a highly nonlinear resistance/temperature response useful for thermal or circuit control, not for temperature measurement. Besides circuit elements used to limit current, self-limiting heaters can be made in the form of wires or strips, useful for heat tracing. PTC thermistors 'latch' into a hot / high resistance state: once hot, they stay in that high resistance state, until cooled. In fact, Neil A Downie showed how you can use the effect as a simple latch/memory circuit, the effect being enhanced by using two PTC thermistors in series, with thermistor A cool, thermistor B hot, or vice versa.

Self-heating Effects

When a current flows through a thermistor, it will generate heat which will raise the temperature of the thermistor above that of its environment. If the thermistor is being used to measure the temperature of the environment, this electrical heating may introduce a significant error if a correction is not made. Alternatively, this effect itself can be exploited. It can, for example, make a sensitive air-flow device employed in a sailplane rate-of-climb instrument, the electronic variometer, or serve as a timer for a relay as was formerly done in telephone exchanges.

The electrical power input to the thermistor is just:

$$P_E = IV$$

where I is current and V is the voltage drop across the thermistor. This power is converted to heat, and this heat energy is transferred to the surrounding environment. The rate of transfer is well described by Newton's law of cooling:

$$P_T = K(T(R) - T_0)$$

where $T(R)$ is the temperature of the thermistor as a function of its resistance R, T_0 is the temperature of the surroundings, and K is the dissipation constant, usually expressed in units of milliwatts per degree Celsius. At equilibrium, the two rates must be equal.

$$P_E = P_T$$

The current and voltage across the thermistor will depend on the particular circuit configuration. As a simple example, if the voltage across the thermistor is held fixed, then by Ohm's Law we have $I = V / R$ and the equilibrium equation can be solved for the ambient temperature as a function of the measured resistance of the thermistor:

$$T_0 = T(R) - \frac{V^2}{KR}$$

The dissipation constant is a measure of the thermal connection of the thermistor to its surroundings. It is generally given for the thermistor in still air, and in well-stirred oil. Typical values for a small glass bead thermistor are 1.5 mW/°C in still air and 6.0 mW/°C in stirred oil. If the temperature of the environment is known beforehand, then a thermistor may be used to measure the value of the dissipation constant. For example, the thermistor may be used as a flow rate sensor, since the dissipation constant increases with the rate of flow of a fluid past the thermistor.

The power dissipated in a thermistor is typically maintained at a very low level to ensure insignificant temperature measurement error due to self heating. However, some thermistor applications depend upon significant "self heating" to raise the body temperature of the thermistor well above the ambient temperature so the sensor then detects even subtle changes in the thermal conductivity of the environment. Some of these applications include liquid level detection, liquid flow measurement and air flow measurement.

Applications

PTC

- As current-limiting devices for circuit protection, as replacements for fuses. Current through the device causes a small amount of resistive heating. If the current is large enough to generate more heat than the device can lose to its surroundings, the device heats up, causing its resistance to increase. This creates a self-reinforcing effect that drives the resistance upwards, therefore limiting the current.
- As timers in the degaussing coil circuit of most CRT displays. When the display unit is initially switched on, current flows through the thermistor and degaussing coil. The coil and

thermistor are intentionally sized so that the current flow will heat the thermistor to the point that the degaussing coil shuts off in under a second. For effective degaussing, it is necessary that the magnitude of the alternating magnetic field produced by the degaussing coil decreases smoothly and continuously, rather than sharply switching off or decreasing in steps; the PTC thermistor accomplishes this naturally as it heats up. A degaussing circuit using a PTC thermistor is simple, reliable (for its simplicity), and inexpensive.

- As heater in automotive industry to provide additional heat inside cabin with diesel engine or to heat diesel in cold climatic conditions before engine injection.
- In temperature compensated synthesizer voltage controlled oscillators.
- In lithium battery protection circuits.
- In an electrically actuated Wax motor to provide the heat necessary to expand the wax.

NTC

- As a resistance thermometer for low-temperature measurements of the order of 10 K.
- As an inrush current limiter device in power supply circuits, they present a higher resistance initially, which prevents large currents from flowing at turn-on, and then heat up and become much lower resistance to allow higher current flow during normal operation. These thermistors are usually much larger than measuring type thermistors, and are purposely designed for this application.
- As sensors in automotive applications to monitor things like coolant or oil temperature inside the engine, and provide data to the ECU and to the dashboard.
- To monitor the temperature of an incubator.
- Thermistors are also commonly used in modern digital thermostats and to monitor the temperature of battery packs while charging.
- Thermistors are often used in the hot ends of 3D printers; they monitor the heat produced and allow the printer's control circuitry to keep a constant temperature for melting the plastic filament.
- In the food handling and processing industry, especially for food storage systems and food preparation. Maintaining the correct temperature is critical to prevent foodborne illness.
- Throughout the consumer appliance industry for measuring temperature. Toasters, coffee makers, refrigerators, freezers, hair dryers, etc. all rely on thermistors for proper temperature control.
- NTC thermistors come in bare and lugged forms, the former is for point sensing to achieve high accuracy for specific points, such as laser diode die, etc.

History

The first NTC thermistor was discovered in 1833 by Michael Faraday, who reported on the semiconducting behavior of silver sulfide. Faraday noticed that the resistance of silver sulfide decreased dramatically as temperature increased. (This was also the first documented observation of a semi-conducting material.)

Because early thermistors were difficult to produce and applications for the technology were limited, commercial production of thermistors did not begin until the 1930s. A commercially viable thermistor was invented by Samuel Ruben in 1930.

Semiconductor Sensors (Thermistors)

Thermistors are temperature sensitive variable resistor made of semi-conductor material. Thermistors are made of metal oxides and their mixtures viz. Oxides of Copper, Nickel, Manganese, Iron, Tin, etc.

They are available in

- beads as small as 0.4mm in diameter
- discs ranging from 5 to 25mm diameter
- rods (a few mm diameter 50mm long)

The temperature resistance relation of thermistors is given by the equation:

$$R \quad R\,e$$

$$= \left(\frac{1}{T} - \frac{1}{T}\right)$$

where R_T is resistance at temperature T

R_O is resistance at T_O

β is a constant ,

β has a value between 3400 and 4000 depending on the composition of the thermistor.

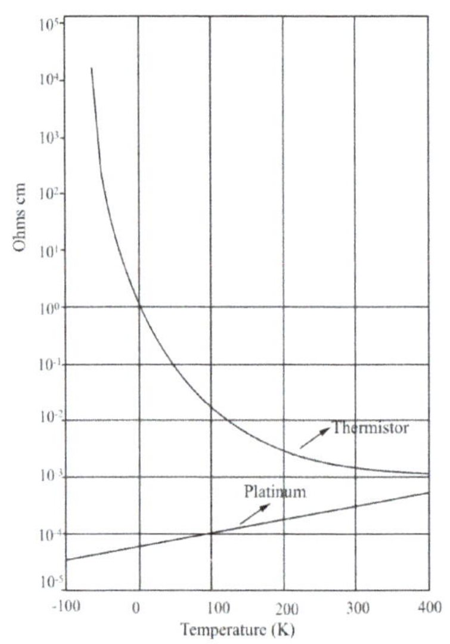

Temperature resistance relation of resistance thermometers

Multiple thermistor sensors can cover a temperature range of -200°C to +1000°C.

Because of the high sensitivity at the lower temperature range thermistors are more commonly used in the lower temperature range.

Thermoelectric Thermometry

An e.m.f is generated, when junctions of two dissimilar metals are kept at different temperatures. The combination of the two metals is called thermocouple.

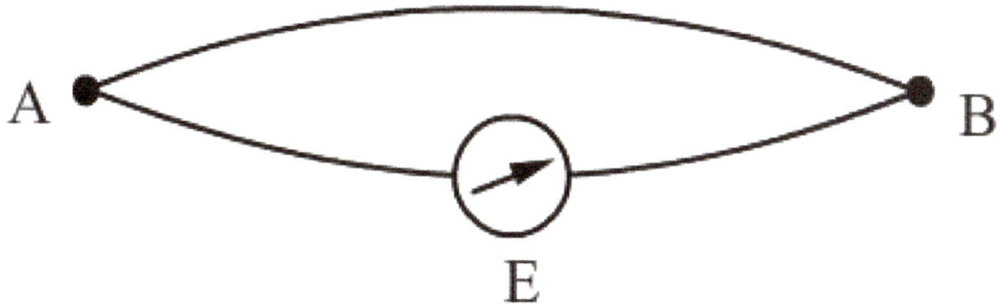

The Seebeck set up

The magnitude of e.m.f depends on the difference in temperature between the two junctions. This is called the Seebeck effect. For a combination of metals A and B as in Fig., the Seebeck voltage dE_S for very small temperature difference is

$$dE_S = \alpha_{A,B} dT$$

$\alpha_{A,B} \rightarrow$ Seebeck coefficient

Laws of Thermocouples

Law of Homogeneous Circuit

A thermo electric current cannot be sustained in a circuit of a single homogeneous material however varying in c.s. by the application of heat alone. The implication is that two different materials are needed to form a thermocouple.

Law of Intermediate Materials

Insertion of an intermediate metal into a thermocouple circuit will not effect the net emf provided the two junctions introduced by the intermediate metal are at identical temperature. This means that there can be a measuring instrument, soldered or brazed between the two metals in order to monitor the emf generated.

Law of Intermediate Temperature

If a thermocouple develops an e.m.f e1 when the junctions are at T_1 and T_2 and an e.m.f e_2 when the junctions are at T_2 and T_3, it will develop an e.m.f $e_1 + e_2$ when the junctions are at T_1 and T_3.

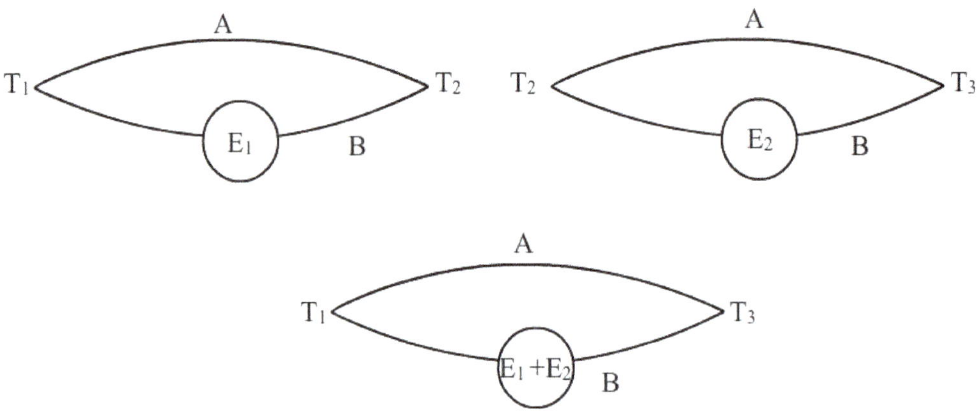

Law of intermediate temperature

$$20 - 100 \simeq 20mv$$

$$100 - 200 \simeq 18mv$$

$$20 - 200 \simeq 38mv$$

Thermocouple Materials

Any two conducting materials could be used as thermocouples. But, certain metals are found to be better than others.

Material	Temperature range
Copper - Constantan	-200 to + 350 ^{0}C
Chromel - Alumel	-200 to 1300 ^{0}C
Iron - Constantan	-150 to 1000 ^{0} C
Pt – (Pt - 10) Rh Pt – (Pt - 13) Rh	0 ^{0}C to 1450 ^{0} C

Thermopiles and Thermocouples Connected in Parallel

Thermocouples may be computed electrically series or parallel. When connected in series, the combination is thermopile.

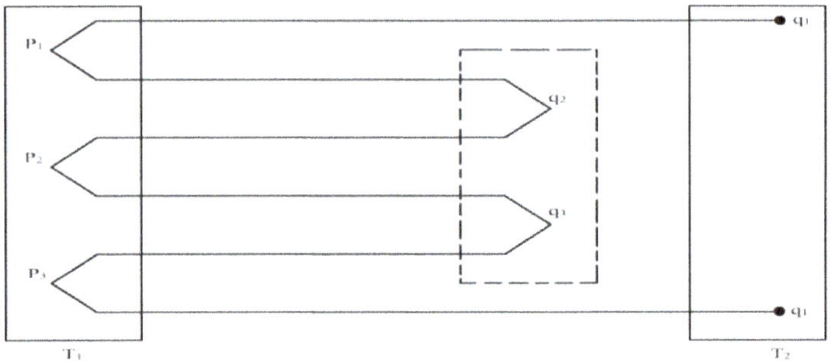

Thermocouples connected in series – thermopile

Here P_1, P_2 and P_3 are the measurement points and Q_1 to Q_4 are the reference points. The total output from n thermocouples will be equal to the sum of individual emfs. So the purpose is to get a more sensitive measurement.

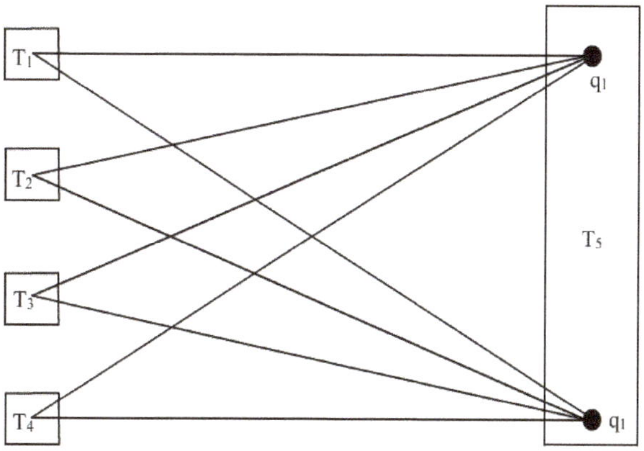

Thermocouples connected in parallel

Thermocouples when connected in parallel connection provides better averaging. The parallel combination gives the same voltage if all the measuring and reference junctions are at the same temperature, If all the measuring junctions are at different temperatures and the thermocouples have the same properties, the voltage measured is the average of the individual voltages.

Pyrometry

The word is derived from pyros + metron. The methods under this are primarily thermal radiation measurement. There are two distinct instruments. Under this category:

(a) Total Radiation Pyrometer

(b) Optical pyrometer

Total Radiation Pyrometer

Total radiation pyrometer accepts a controlled sample of total radiation and through determination of the heating effect of the sample obtains a measure of temperature. All bodies above absolute zero temperature radiate energy, not only do they radiate or emit energy, but they also receive and absorb from other sources. It is known that all substances emit and absorb radiant energy at a rate depending on the absolute temperature and physical properties of the substance.

Stefan-Boltzman Law

According to Stefan – Boltzman law the net rate of exchange of energy between two ideal radiators A and B is,

$$q = \sigma\left(T_A^4 - T_B^4\right)$$

$q \rightarrow$ Radiative heat transfer

$\sigma \rightarrow$ Stefan-Boltzman Constant

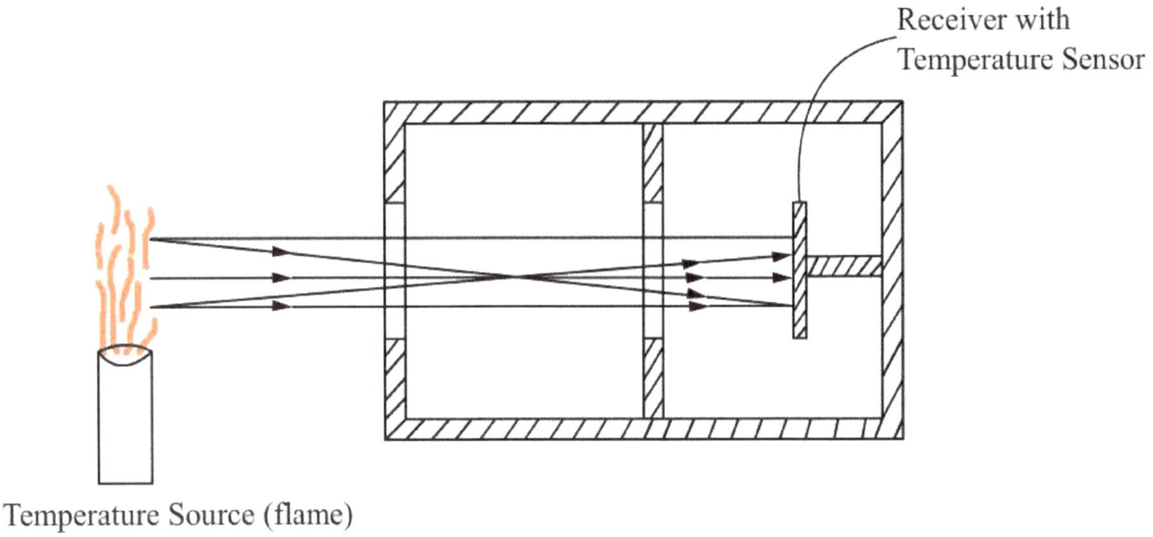

Total radiation pyrometer

In total radiation pyrometers, the radiation from the measured body is focused on some sort of radiation detector which produces an electric signal. Detectors may be classified as thermal detectors or photon detectors. Thermal detectors are blackened elements designed to absorb a maximum of the incoming radiation at all wavelengths. The absorbed radiation causes the temperature of the detector to rise until equilibrium is reached with heat losses to the surroundings. The thermal detectors measure this temperature using a resistance thermometer, thermistor or thermocouple.

In photon detectors, the incoming radiation frees electrons in the detector structure and produces a measurable electrical effect. These events occur on an atomic or molecular time scale and hence are faster than the thermal detectors. However, photon detectors have a sensitivity that varies with wave length, thus incoming radiation of all wavelengths are not equally treated.

Optical Pyrometer

Optical pyrometer employs an optical means for estimating the change in average wavelength of visual radiation with temperature. The instrument works on the principle of Wien's displacement law states that the wavelength distribution of thermal radiation from a black body at any temperature has essentially the same shape as the distribution at any other temperature, except that each wavelength is displaced. There is a shift in the wavelength of maximum emission toward shorter waves. (From Red to blue).when the temperature increases. The intensity relation is expressed as

$$E_\lambda = C_1\lambda^{-5} / \left[e^{(C_2/\lambda T)} - 1\right]$$

E_λ = energy emitted at Wavelength λ

C_1 and C_2 = constants

T = absolute temperature of blackbody

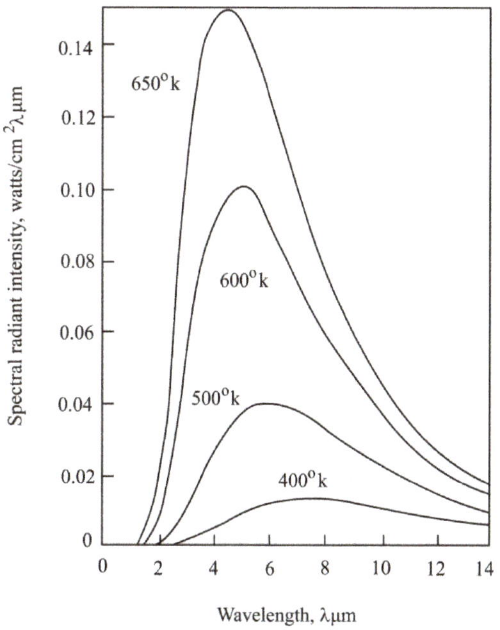

Representation of the Wien's displacement law

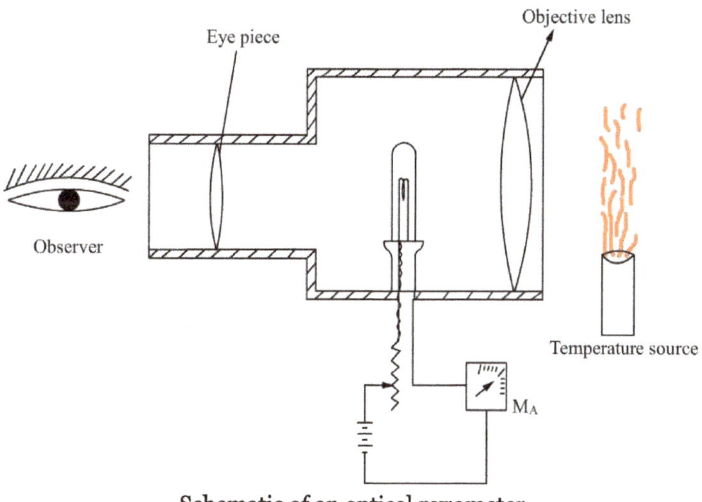

Schematic of an optical pyrometer

In the operation of the pyrometer, method of matching is used. Reference temperature is obtained by an electrically heated filament lamp which is controllable. A measure of temperature is obtained by optically comparing the visual radiation from filament with that from the unknown source.

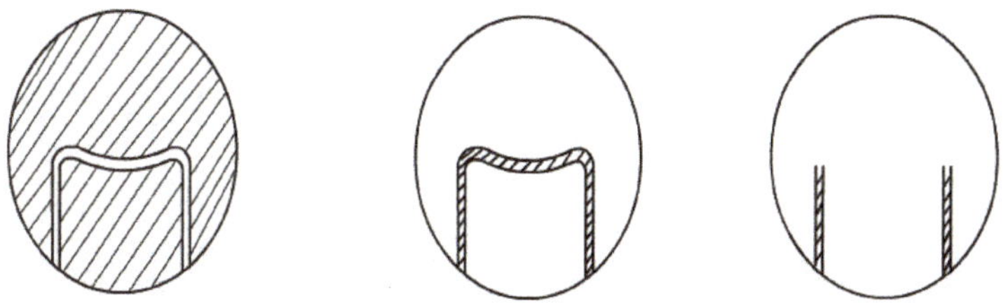

Filament too bright	Filament too cold	Filament and source at equal temperature

The heated filament and the temperature source at different temperatures

Measurement of Temperature in Flow

Measurement of the temperature of flowing gas is important in many practical cases. The state of the stationary perfect gas can be defined by two independent physical parameters one of which may be the temperature. When velocity of flow is such that compressibility effects are important, it is necessary to differentiate between static and stagnation temperatures.

Stagnation temperature is that reached by the fluid when it is brought to rest adiabatically.

A thermometer moving with the fluid and emitting no thermal radiation would measure static temperature. An intrusive sensor cannot measure static temperature. Static temperature is measured by non-intrusive methods or estimated indirectly. (by measuring static pressure and then density by optical means) or by measuring the speed of sound. The difference between stagnation temperature T_O and free temperature T if a moving perfect gas can be determined from

$$T_0 - T = \frac{V^2}{2C_p}$$

Since, shocks do not change the enthalpy; this equation is true for both subsonic and supersonic flows.

Adiabatic Wall Temperature or Recovery Temperature

A thermally insulated surface will be heated by a gas flowing past it to a temperature called recovery temperature (T_a). The recovery temperature depends on

- Local mach number (or on static temperature) at the outside limit of boundary layer.
- On the dissipation of kinetic energy by friction in the boundary layer.
- On the rate of heat exchange.

$$T_a - T = K\frac{V^2}{2C_p}$$

Difference between recovery temperature and static temperature is a fraction of the adiabatic temperature rise.

$$K = \frac{T_a - T}{T_0 - T}$$

K is called the coefficient of thermal recovery or the recovery coefficient. K represents the proportion of kinetic energy of the medium recovered as heat. K depends on the shape of the body and on Mach number, Reynolds number, Prandtl number and on the ratio of specific heats γ.

For a given gas Pr and γ are constant over a wide range of temperatures (For air Pr = 0.72 and γ = 1.4) and so recovery coefficient is a function of M and Re only, where, k is the coefficient of thermal conductivity $P_r = \frac{\mu C_p}{k}$.

The coefficient of thermal recovery depends on the shape of the surface. For poorly streamlined bodies r varies between 0.6 and 0.7 and for well streamlined bodies it is between

0.8 and 0.9. The relationship between recovery temperature and stagnation temperature depends on the Mach number and can be derived from following two equations:

$$\frac{T_0}{T} = 1 + \frac{\gamma - 1}{2} M^2 \text{ and}$$

$$\gamma = \frac{T_a - T}{T_0 - T}$$

$$\frac{T_a}{T_0} = 1 - \frac{\frac{\gamma - 1}{2} M^2}{1 + \frac{\gamma - 1}{2} M^2} (1 - \mu)$$

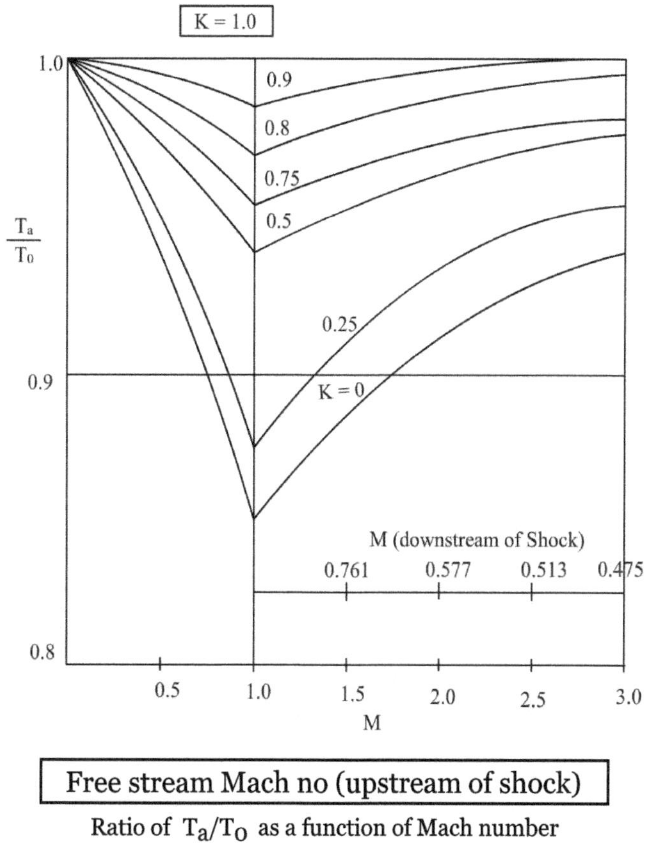

Ratio of T_a/T_0 as a function of Mach number

When M exceeds unity a shock appears whose strength increases with increasing incident Mach number. In the absence of heat transfer a thermometer on the wall of a tube inserted into a gas stream would indicate a recovery temperature T_a dependent only on the flow characteristics in the boundary layer around the tube. When, K = 1, $T_a = T_0$.

In an actual thermometer of a temperature probe, heat exchange with surrounding medium cannot be prevented. Hence, it will indicate a temperature T_n differentiating from the recovery temperature T_a. Recovery coefficient of the instrument

$$K = \frac{T_n - T}{T_0 - T}$$

The value of K can be found by calibrating. With the recovery coefficient known and knowing T_n as indicated by the instrument, T_0 can be found by substituting in

$$\frac{T_n}{T_0} = 1 - \frac{\frac{\gamma - 1}{2} M^2}{1 + \frac{\gamma - 1}{2} M^2}(1 - K)$$

Temperature Measurement Problems in Flow

Conduction Errors

When temperature is measured with a probe, heat is transferred from probe to environment. A probe is inserted into the flow and is supported at the wall. In general the wall is either hotter or colder than the fluid. Therefore, probe temperature is different from that of the fluid. Fluid at T_f exchanges heat with the probe by convection. A simplified model is assumed to find what measures can be taken to reduce error in such a case.

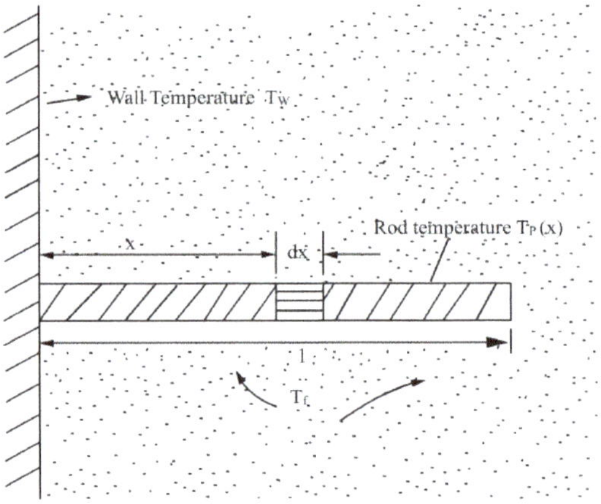

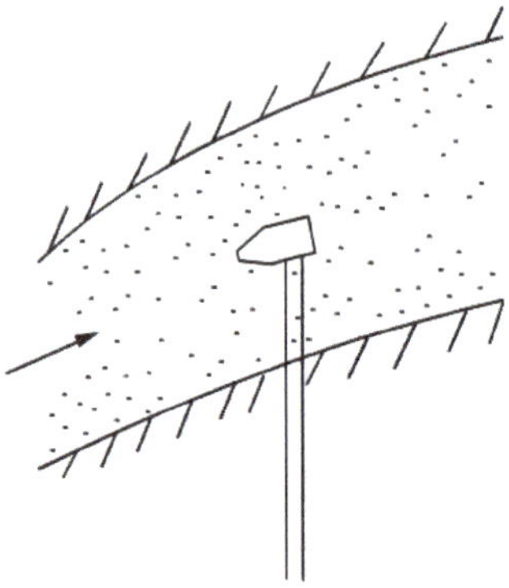

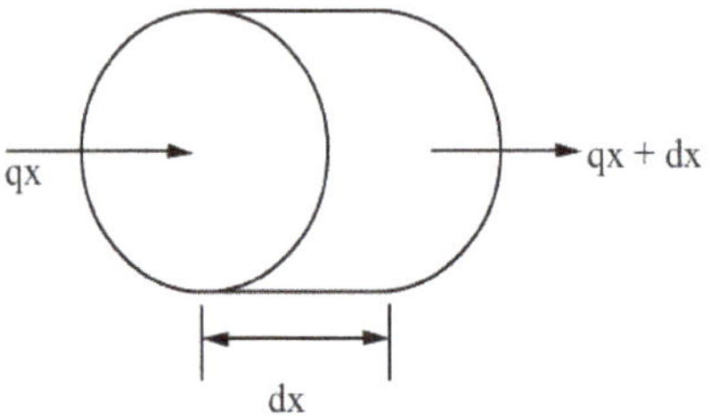

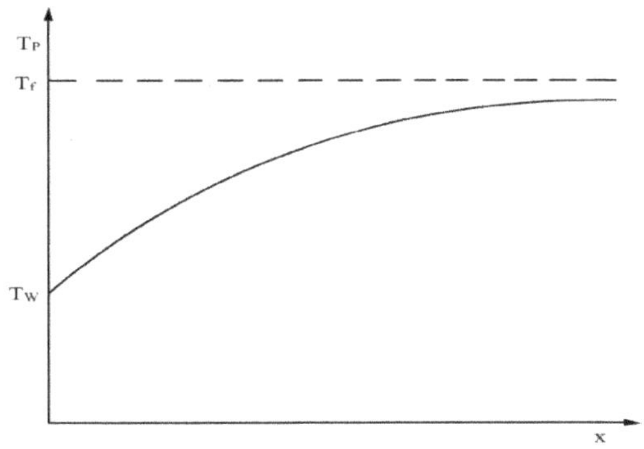

Probe configuration in a flow

Assuming steady state situation, heat in at x = (heat out at x+dx) +heat loss at surface.

Assumed that rod temperature is a function of x only. It does not change with time or over rod cross section at a given x.

$$q_x = q_{x+dx} + q_l$$

One - dimensional conduction heat transfer gives

$$q_x = -kA\frac{dT_p}{dx}$$

where k is thermal conductivity of probe or rod. A is its cross sectional area. T_p is temperature of the probe

$$= -kA\frac{dT_p}{dx} + \frac{d}{dx}\left(-kA\frac{dT_p}{dx}\right)dx$$

$$= -kA\frac{dT}{dx} + \frac{d}{dx}\left(-kA\frac{dT_p}{dx}\right)dx$$

k and A are assumed constants

$$qx + dx = -kA\frac{dT_p}{dx} - kA\frac{d^2T_p}{dx^2}dx$$

Heat loss by convection at surface

$$q_l = h(Cdx)(T_p - T_f)$$

where, h = film coefficient of heat transfer

C = Circumference of rod

C dx =Surface area

So, we have

$$\frac{d^2T_p}{dx^2} - \frac{hC}{kA}T_p = -\frac{hC}{kA}T_f$$

This is a linear differential equation

If h and C are constants, we have a linear differential equation with constant coefficients and is solved for T_P as a function of x. Two boundary conditions are needed.

i. $T_p = T_w \; at \; x = 0$

ii. If end is assumed to be insulated

$$\frac{dT_P}{dx} = 0 \;\; at \;\; x = L$$

iii. Even if this is not assumed, from the temperature distribution

$$\frac{dT_P}{dx} = 0 \;\; at \;\; x = L$$

Using these boundary conditions

$$T_p = T_f - (T_f - T_W)\left[\left(1 - \frac{e^{mL}}{2conshmL}\right)e^{mx} + \frac{e^{mL}}{2conshmL}e^{-mx}\right]$$

where, m = $\sqrt{\frac{hC}{kA}}$

Generally the sensor is at x = L

∴ Evaluating we get

Temperature Error $= Tp - Tf = \frac{T_w - T_f}{\cosh mL}$ $\quad (m = \sqrt{\frac{hC}{kA}})$

Error is small if T_W is close to T_f and if coshmL is large hence to reduce the error, insulate the wall cosh mL will be large if m and L are large. So, the probe should be inserted deeper. To make m large, h should be large (high rate of convectional heat transfer) k should be small. To achieve this support the probe with an insulating material $\frac{C}{A}$ depends on shape of the rod. For circular $\frac{C}{A} = \frac{2\pi r}{\pi r^2} = \frac{2}{r}$ where r is the radius.

Radiation Errors

This occurs simultaneously with conduction losses. Only radiation exchange between probe and surroundings is considered now. Neglecting conduction, for steady conditions

Heat convected to probe = net heat radiated to wall.

$$hA_S(T_f - T_P) = \epsilon_P A_S\left[(T_P)^4 - (T_w)^4\right]$$

A_S = surface area of the probe

T_P = probe absolute temperature

$$Nu = \frac{hl}{k}, Nu = f(\Pr, \mathrm{Re})$$

$$Nu = 0.332 \Pr^{\frac{1}{3}} \mathrm{Re}^{\frac{1}{2}}$$

$$\therefore T_P - T_f = \frac{\epsilon_P}{h}\left[T_w^4 - T_P^4\right]$$

ϵ_p includes the emittance of the body.

To reduce radiation error,

a) insulate the wall (ie. reduce T_W - T_P)

b) Use surface of law emittance (shining surface) [ϵ_P to be as small as possible]

c) To obtain high value of h, aspirated type of probe can be used.

(Connect a vacuum pump to the probe and this create high local velocity)

Velocity Effects on Temperature Measurements

Real temperature probes do not attain the theoretical stagnation temperature. Even if conduction and radiation errors are corrected for there remains deviation of actual situation from ideal. Correction for these effects generally is accomplished by experimental calibration to determine recovery factor K of the probe.

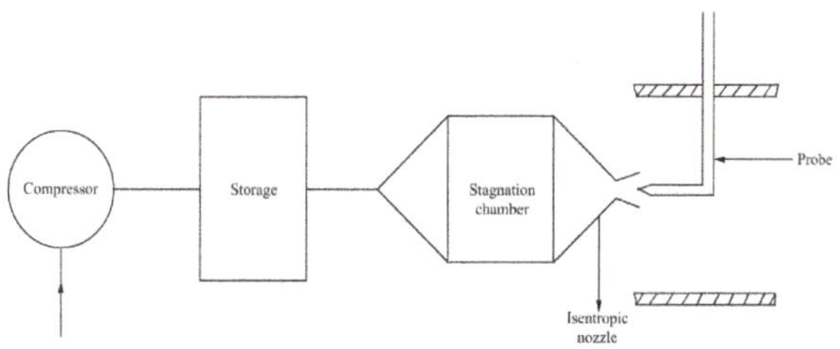

Fig.6.15 Set up for calibration of recovery coefficient

Stagnation chamber velocity $= \frac{1}{100}$ nozzle flow velocity.

Measurement of P_O T_O can be found under zero-velocity conditions.

$T_{stag, nozzle} = T_{stat, tank}$

$P_{stag, nozzle} = P_{stat, tank}$

$P_{stat, nozzle} = P_{atmosphere}$

No pitot tube is used.

$P_{stag, nozzle}$ is got by $P_{stat, tank}$ and $P_{stat, nozzle}$ is obtained by a barometer.

$$T_{stat\ nozzle} = \frac{T_{stag\ nozzle}}{1+\left(\frac{\gamma-1}{2}\right)M^2} = \frac{T_{\text{stat tank}}}{1+\left(\frac{\gamma-1}{2}\right)M^2}$$

Once, M is known $T_{\text{stat nozzle}}$ can be found.

The probe reading gives, T_a or the indicated temperature

$$K = \frac{T_a - T_{\text{stat,nozzle}}}{T_{\text{stag,nozzle}} - T_{\text{stat,nozzle}}}$$

Sensors / Probes for Measuring Stagnation Temperature

Sensor depends on the intended range of flow velocities and temperatures. For a good sensor

(1) the value of K should be as close to unity (2) the value should be constant. The deviation from unity of the value K depends on

(1) Convectional heat exchange between sensing element and medium

(2) Heat loss by conduction from the sensor through the device holding it.

(3) Radiant heat exchange between sensor and surroundings.

Since, the above factors depend on temperature and velocity. Probes are divided into a few categories.

- Sensors for low and high velocities at low temperature.
- Sensors for high velocities and temperature upto 300 and 400°C
- Sensors for low and high velocities (upto 1000 to 1200)

a) Low temperature sensors

Losses due to radiation can be neglected when the probe and walls of the flow channel do not differ much in temperature. To determine the flow temperature, stagnation temperature in the settling chamber can be determined settling chamber temperature can be determined where there is no velocity. As there is no addition or removal of heat between the settling chamber and the test section, the settling chamber temperature can be assumed to be same as the test section temperature. Mercury thermometers, resistance thermometers or thermocouples could be used as the sensor.

b) Sensors for high velocities and medium temperature

For temperature range from 300 to 400°C, the sensors are mounted in narrow channels. Thermocouple wires of 0.1 to 0.2mm diameter [iron =constantan or copper constantan] t are generally employed in this range of temperature. The thermal capacity of the junction of thermocouple is very small so that it responds rapidly and measurement can be done at rapidly changing temperatures. When there is no major radiant heat transfer, a thermocouple consisting of the thermocouple junction inserted lengthwise into the flowing medium will have a recovery coefficient of ~0.9 even up to sonic velocities.

Well made stagnation temperature probes will have recovery coefficients close to unity over a wide range of velocities. The gas upstream of the sensor is brought to an optimum velocity such that heat gained by the sensor due to convection will be balanced by heat lost by conduction. In order that the flow is not brought to rest completely, the tube is provided with vents of area 1/4 to 1/8 of the inlet orifice area.

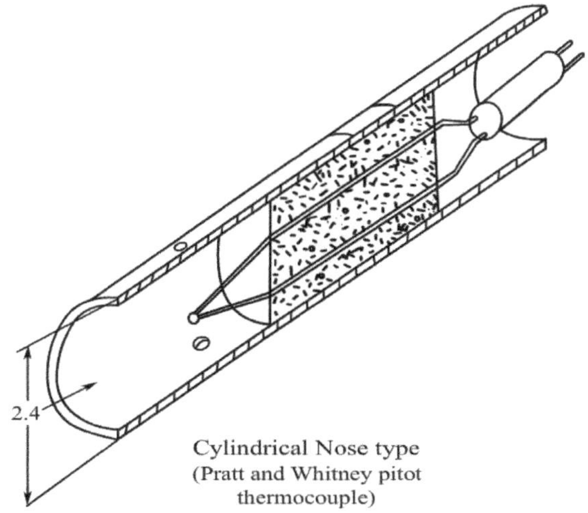

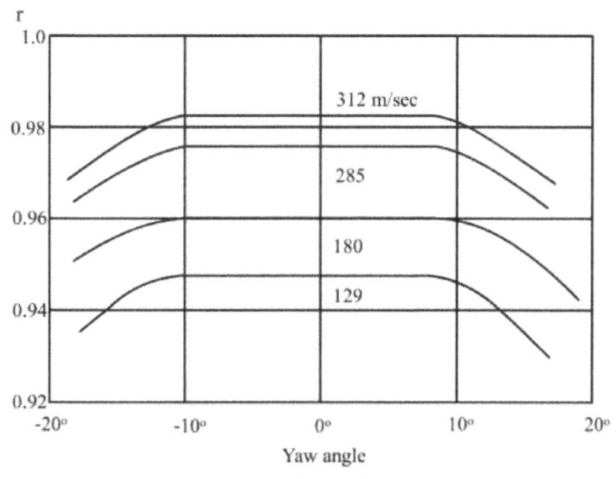

Total temperature probe and its characteristics

Figure above shows one such probe. Such probes are simple to make and have recovery coefficient between 0.95 and 0.99.

c) High temperature probes (for temperatures above 300 to 400°C

In such cases, the temperature difference between surrounding medium and sensor is 50°C. Hence radiant heat losses become predominant. Radiation intensity depends on area and because of this reason high temperature sensors are made very small. The radiation capacity of the body on which the sensor is mounted should be low and this is achieved by polishing. Another way to achieve good results is by shielding the sensor using concentric tubes.

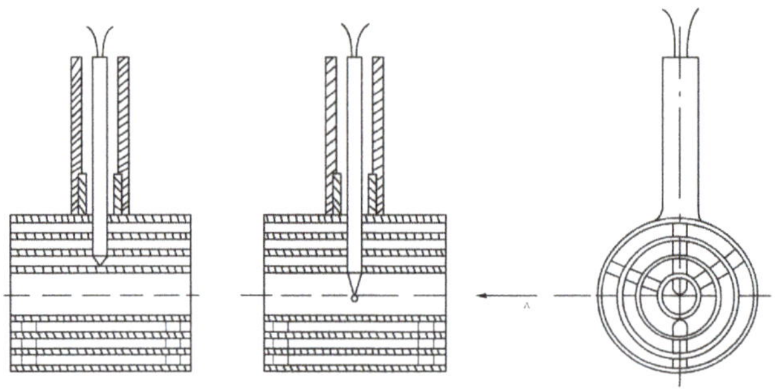

Shielded sensor for high temperatures

Temperature probe with concentric tubes

The sensor may be welded to the innermost tube or in the centre of the inner tube. In the first case it will act as a poorly streamlined body and the recovery coefficient will be ~0.65. In the latter case the recovery coefficient may be up to 0.9.The innermost tube may be of insulator material (porcelain) and the outer tubes of heat resistant steel.

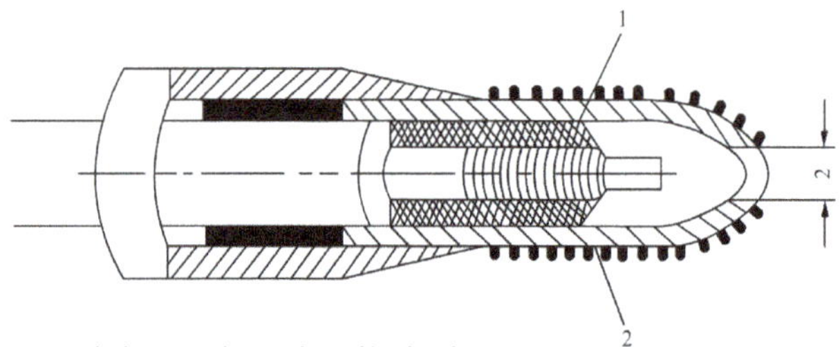

1- Main thermocouple mounting and heating element
2 - Radiation shield and heating element

Temperature probe with heated shield

Radiation losses can be reduced by heating the shield to a temperature close to the ambient temperature of the medium. Figure above gives a miniature probe with heated shied developed by California Institute of Technology. In this an electrically heated wire on the shield reduces direct radiation losses and by conduction from the shield. To reduce the heat loss by conduction through the leads from the thermocouple and its holder, the latter is also heated by electric heating.

Pyrometer

An optical pyrometer

A sailor checking the temperature of a ventilation system.

A pyrometer is a type of remote-sensing thermometer used to measure the temperature of a surface. Various forms of pyrometers have historically existed. In the modern usage, it is a device that from a distance determines the temperature of a surface from the spectrum of the thermal radiation it emits, a process known as pyrometry and sometimes radiometry.

The word pyrometer comes from the Greek word for fire, (*pyro*), and *meter*, meaning to measure. The word pyrometer was originally coined to denote a device capable of measuring the temperature of an object by its incandescence, visible light emitted by a body which is at least red-hot. Modern pyrometers or infrared thermometers also measure the temperature of cooler objects, down to room temperature, by detecting their infrared radiation flux.

Design

A modern pyrometer has an optical system and a detector. The optical system focuses the thermal radiation onto the detector. The output signal of the detector (temperature T) is related to the thermal radiation or irradiance j^* of the target object through the Stefan–Boltzmann law, the constant of proportionality σ, called the Stefan-Boltzmann constant and the emissivity ε of the object.

$$j^{\star} = \varepsilon\sigma T^4$$

This output is used to infer the object's temperature from a distance, with no need for the pyrometer to be in thermal contact with the object; most other thermometers (e.g. thermocouples and resistance temperature detectors (RTDs)) are placed in thermal contact with the object, and allowed to reach thermal equilibrium.

Pyrometry of gases presents difficulties. These are most commonly overcome by using thin filament pyrometry or soot pyrometry. Both techniques involve small solids in contact with hot gases.

History

A pyrometer from 1852. Heating the metal bar (a) presses against a lever (b), which moves a pointer (c) along a scale that serves as a measuring index. (e) is an immovable prop which holds the bar in place. A spring on (c) pushes against (b), causing the index to fall back once the bar cools.

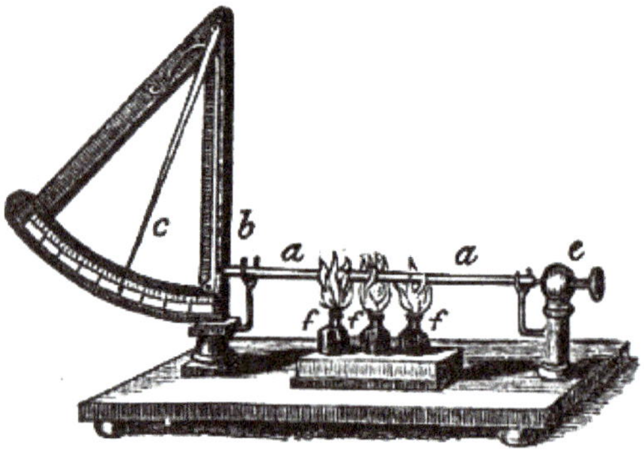

The potter Josiah Wedgwood invented the first pyrometer to measure the temperature in his kilns, which first compared the color of clay fired at known temperatures, but was eventually upgraded to measuring the shrinkage of pieces of clay, which depended on kiln temperature.

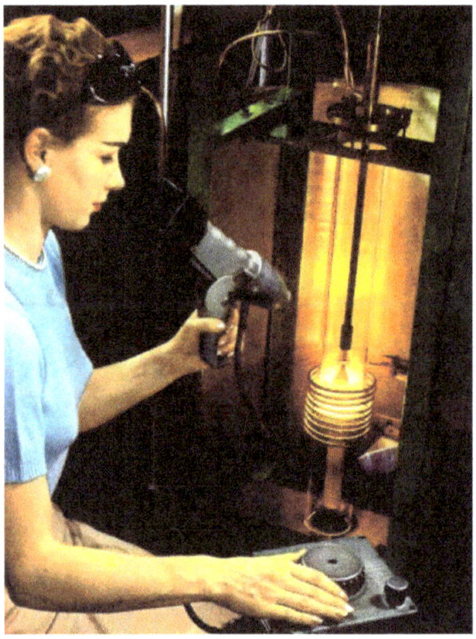

Technician measuring the temperature of molten silicon at 2650°F with a disappearing filament pyrometer in Czochralski crystal growing equipment at Raytheon transistor plant in 1956.

The first disappearing filament pyrometer was built by L. Holborn and F. Kurlbaum in 1901. This device had a thin electrical filament between an observer's eye and an incandescent object. The current through the filament was adjusted until it was of the same colour (and hence temperature) as the object, and no longer visible; it was calibrated to allow temperature to be inferred from the current.

The temperature returned by the vanishing filament pyrometer and others of its kind, called brightness pyrometers, is dependent on the emissivity of the object. With greater use of brightness pyrometers, it became obvious that problems existed with relying on knowledge of the value of emissivity. Emissivity was found to change, often drastically, with surface roughness, bulk and surface composition, and even the temperature itself.

To get around these difficulties, the ratio or two-color pyrometer was developed. They rely on the fact that Planck's law, which relates temperature to the intensity of radiation emitted at individual wavelengths, can be solved for temperature if Planck's statement of the intensities at two different wavelengths is divided. This solution assumes that the emissivity is the same at both wavelengths and cancels out in the division. This is known as the gray body assumption. Ratio pyrometers are essentially two brightness pyrometers in a single instrument. The operational principles of the ratio pyrometers were developed in the 1920s and 1930s, and they were commercially available in 1939.

As the ratio pyrometer came into popular use, it was determined that many materials, of which metals are an example, do not have the same emissivity at two wavelengths. For these materials, the emissivity does not cancel out and the temperature measurement is in error. The amount of error depends on the emissivities and the wavelengths where the measurements are taken. Two-color ratio pyrometers cannot measure whether a material's emissivity is wavelength dependent.

To more accurately measure the temperature of real objects with unknown or changing emissivities, multiwavelength pyrometers were envisioned at the US National Institute of Standards and Technology and described in 1992. Multiwavelength pyrometers use three or more wavelengths and mathematical manipulation of the results to attempt to achieve accurate temperature measurement even when the emissivity is unknown, changing, and different at all wavelengths.

Applications

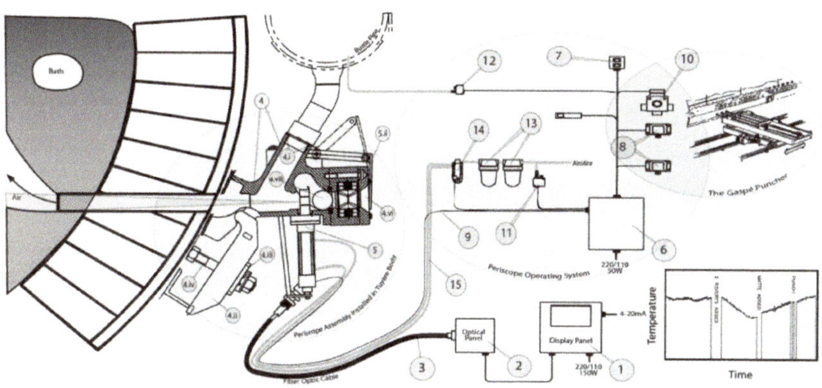

A tuyère pyrometer. (1) Display. (2) Optical. (3) Fibre optic cable and periscope. (4) Pyrometer tuyère adapter having: i. Bustle pipe connection. ii. Tuyère clamp. iii. Clamp washer. iv. Clamp stud c/w and fastening hardware. v. Gasket. vi. Noranda tuyère silencer. vii. Valve seat. viii. Ball. (5) Pneumatic cylinder: i. Smart cylinder assembly with Internal proximity switch. ii. Guard plate assembly. iii. Temporary flange cover plate, used to cover periscope entry hole on tuyère adapter when no cylinder is installed on the tuyère. (6) Operator station panel. (7) Pyrometer light station. (8) Limit switches. (9) 4 conductor cab tire. (10) Ball Valve. (11) Periscope air pressure switch. (12) Bustle pipe air pressure switch. (13) Airline filter/regulator. (14) Directional control valve, sub-plate, silencer and speed control mufflers. (15) 2" nom. low pressure air hose, 40m length.

Pyrometers are suited especially to the measurement of moving objects or any surfaces that can not be reached or can not be touched.

Temperature is a fundamental parameter in metallurgical furnace operations. Reliable and continuous measurement of the melt temperature is essential for effective control of the operation. Smelting rates can be maximized, slag can be produced at the optimum temperature, fuel consumption is minimized and refractory life may also be lengthened. Thermocouples were the traditional devices used for this purpose, but they are unsuitable for continuous measurement because they melt and degrade.

Measuring the combustion temperature of coke in the blast furnace using an optical pyrometer, Fixed Nitrogen Research Laboratory, 1930.

Salt bath furnaces operate at temperatures up to 1300 °C and are used for heat treatment. At very high working temperatures with intense heat transfer between the molten salt and the steel being treated, precision is maintained by measuring the temperature of the molten salt. Most errors are caused by slag on the surface which is cooler than the salt bath.

The *tuyère pyrometer* is an optical instrument for temperature measurement through the tuyeres which are normally used for feeding air or reactants into the bath of the furnace.

A steam boiler may be fitted with a pyrometer to measure the steam temperature in the superheater.

A hot air balloon is equipped with a pyrometer for measuring the temperature at the top of the envelope in order to prevent overheating of the fabric.

Pyrometers may be fitted to experimental gas turbine engines to measure the surface temperature of turbine blades. Such pyrometers can be paired with a tachometer to tie the pyrometer output with the position of an individual turbine blade. Timing combined with a radial position encoder allows engineers to determine the temperature at exact points on blades moving past the probe.

Stefan–Boltzmann Law

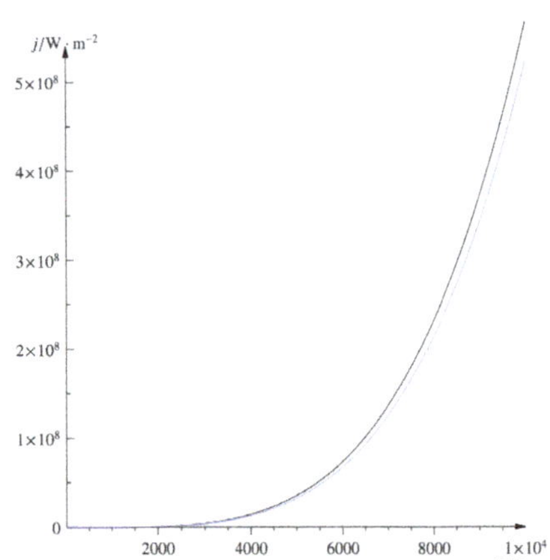

Graph of a function of total emitted energy of a black body $j^{\star}$ proportional to its thermodynamic temperature T. In blue is a total energy according to the Wien approximation, $j_W^{\star} = j^{\star} / \zeta(4) \approx 0.924\sigma T^4$

The Stefan–Boltzmann law describes the power radiated from a black body in terms of its temperature. Specifically, the Stefan–Boltzmann law states that the total energy radiated per unit surface area of a black body across all wavelengths per unit time (also known as the black-body *radiant emittance* or *radiant exitance*), $j^\star$, is directly proportional to the fourth power of the black body's thermodynamic temperature *T*:

$$j^\star = \sigma T^4.$$

The constant of proportionality σ, called the Stefan–Boltzmann constant derives from other known constants of nature. The value of the constant is

$$\sigma = \frac{2\pi^5 k^4}{15c^2 h^3} = 5.670373 \times 10^{-8}\,\mathrm{W\,m^{-2}K^{-4}},$$

where k is the Boltzmann constant, h is Planck's constant, and c is the speed of light in a vacuum. Thus at 100 K the energy flux is 5.67 W/m², at 1000 K 56,700 W/m², etc. The radiance (watts per square metre per steradian) is given by

$$L = \frac{j^\star}{\pi} = \frac{\sigma}{\pi} T^4.$$

A body that does not absorb all incident radiation (sometimes known as a grey body) emits less total energy than a black body and is characterized by an emissivity, $\varepsilon < 1$:

$$j^\star = \varepsilon \sigma T^4.$$

The irradiance $j^\star$ has dimensions of energy flux (energy per time per area), and the SI units of measure are joules per second per square metre, or equivalently, watts per square metre. The SI unit for absolute temperature *T* is the kelvin. ε is the emissivity of the grey body; if it is a perfect blackbody, $\varepsilon = 1$. In the still more general (and realistic) case, the emissivity depends on the wavelength, $\varepsilon = \varepsilon(\lambda)$.

To find the total power radiated from an object, multiply by its surface area, A:

$$P = Aj^\star = A\varepsilon\sigma T^4.$$

Wavelength- and subwavelength-scale particles, metamaterials, and other nanostructures are not subject to ray-optical limits and may be designed to exceed the Stefan–Boltzmann law.

History

The law was deduced by Josef Stefan (1835–1893) in 1879 on the basis of experimental measurements made by John Tyndall and was derived from theoretical considerations, using thermodynamics, by Ludwig Boltzmann (1844–1906) in 1884. Boltzmann considered a certain ideal heat engine with light as a working matter instead of a gas. The law is highly accurate only for ideal black objects, the perfect radiators, called black bodies; it works as a good approximation for most "grey" bodies. Stefan published this law in the article (*On the relationship between thermal radiation and temperature*) in the *Bulletins from the sessions* of the Vienna Academy of Sciences.

Examples

Temperature of the Sun

With his law Stefan also determined the temperature of the Sun's surface. He learned from the data of Charles Soret (1854–1904) that the energy flux density from the Sun is 29 times greater than the energy flux density of a certain warmed metal lamella (a thin plate). A round lamella was placed at such a distance from the measuring device that it would be seen at the same angle as the Sun. Soret estimated the temperature of the lamella to be approximately 1900 °C to 2000 °C. Stefan surmised that ⅓ of the energy flux from the Sun is absorbed by the Earth's atmosphere, so he took for the correct Sun's energy flux a value 3/2 times greater than Soret's value, namely 29 × 3/2 = 43.5.

Precise measurements of atmospheric absorption were not made until 1888 and 1904. The temperature Stefan obtained was a median value of previous ones, 1950 °C and the absolute thermodynamic one 2200 K. As $2.57^4 = 43.5$, it follows from the law that the temperature of the Sun is 2.57 times greater than the temperature of the lamella, so Stefan got a value of 5430 °C or 5700 K (the modern value is 5778 K). This was the first sensible value for the temperature of the Sun. Before this, values ranging from as low as 1800 °C to as high as 13,000,000 °C were claimed. The lower value of 1800 °C was determined by Claude Servais Mathias Pouillet (1790–1868) in 1838 using the Dulong-Petit law. Pouillet also took just half the value of the Sun's correct energy flux.

Temperature of Stars

The temperature of stars other than the Sun can be approximated using a similar means by treating the emitted energy as a black body radiation. So:

$$L = 4\pi R^2 \sigma T_e^4$$

where L is the luminosity, σ is the Stefan–Boltzmann constant, R is the stellar radius and T is the effective temperature. This same formula can be used to compute the approximate radius of a main sequence star relative to the sun:

$$\frac{R}{R_\odot} \approx \left(\frac{T_\odot}{T}\right)^2 \cdot \sqrt{\frac{L}{L_\odot}}$$

where $R_\odot$, is the solar radius, $L_\odot$ is the solar luminosity, and so forth.

With the Stefan–Boltzmann law, astronomers can easily infer the radii of stars. The law is also met in the thermodynamics of black holes in so-called Hawking radiation.

Effective Temperature of the Earth

Similarly we can calculate the effective temperature of the Earth $T_\oplus$ by equating the energy received from the Sun and the energy radiated by the Earth, under the black-body approximation. The luminosity of the Sun, $L_\odot$, is given by:

$$L_\odot = 4\pi R_\odot^2 \sigma T_\odot^4$$

At Earth, this energy is passing through a sphere with a radius of a_0, the distance between the Earth and the Sun, and the irradiance (received power per unit area) is given by

$$E_\oplus = \frac{L_\odot}{4\pi a_0^2}$$

The Earth has a radius of $R_\oplus$, and therefore has a cross-section of $\pi R_\oplus^2$. The radiant flux (i.e. solar power) absorbed by the Earth is thus given by:

$$\Phi_{abs} = \pi R_\oplus^2 \times E_\oplus :$$

Assuming the exchange is in a steady state, the flux emitted by Earth must equal the flux absorbed, and so:

$$4\pi R_\oplus^2 \sigma T_\oplus^4 = \pi R_\oplus^2 \times E_\oplus$$

$$= \pi R_\oplus^2 \times \frac{4\pi R_\odot^2 \sigma T_\odot^4}{4\pi a_0^2}$$

$T_\oplus$ can then be found:

$$T_\oplus^4 = \frac{R_\odot^2 T_\odot^4}{4a_0^2}$$

$$T_\oplus = T_\odot \times \sqrt{\frac{R_\odot}{2a_0}}$$

$$= 5780\ \text{K} \times \sqrt{\frac{696 \times 10^6\ \text{m}}{2 \times 149.598 \times 10^9\ \text{m}}}$$

$$\approx 279$$

where $T_\odot$ is the temperature of the Sun, $R_\odot$ the radius of the Sun, and a_0 is the distance between the Earth and the Sun. This gives an effective temperature of 6 °C on the surface of the Earth, assuming that it perfectly absorbs all emission falling on it and has no atmosphere.

The Earth has an albedo of 0.3, meaning that 30% of the solar radiation that hits the planet gets scattered back into space without absorption. The effect of albedo on temperature can be approximated by assuming that the energy absorbed is multiplied by 0.7, but that the planet still radiates as a black body (the latter by definition of effective temperature, which is what we are calculating). This approximation reduces the temperature by a factor of $0.7^{1/4}$, giving 255 K (−18 °C).

However, long-wave radiation from the surface of the earth is partially absorbed and re-radiated back down by greenhouse gases, namely water vapor, carbon dioxide and methane. Since the emissivity with greenhouse effect (weighted more in the longer wavelengths where the Earth radiates) is reduced more than the absorptivity (weighted more in the shorter wavelengths of the Sun's radiation) is reduced, the equilibrium temperature is higher than the simple black-body calculation estimates. As a result, the Earth's actual average surface temperature is about 288 K (15 °C), which is higher than the 255 K effective temperature, and even higher than the 279 K temperature that a black body would have.

In the above discussion, we have assumed that the whole surface of the earth is at one temperature. Another interesting question is to ask what the temperature of a blackbody surface on the earth would be assuming that it reaches equilibrium with the sunlight falling on it. This of course depends on the angle of the sun on the surface and on how much air the sunlight has gone through. When the sun is at the zenith and the surface is horizontal, the irradiance can be as high as 1120 W/m². The Stefan–Boltzmann law then gives a temperature of

$$T = \left(\frac{1120\ \mathrm{W/m^2}}{\sigma} \right)^{1/4} \approx 375\ \mathrm{K}$$

or 102 °C. (Above the atmosphere, the result is even higher: 394 K.) We can think of the earth's surface as "trying" to reach equilibrium temperature during the day, but being cooled by the atmosphere, and "trying" to reach equilibrium with starlight and possibly moonlight at night, but being warmed by the atmosphere.

Origination

Thermodynamic Derivation of the Energy Density

The fact that the energy density of the box containing radiation is proportional to T^4 can be derived using thermodynamics. It follows from the Maxwell stress tensor of classical electrodynamics that the radiation pressure p is related to the internal energy density u:

$$p = \frac{u}{3}.$$

From the fundamental thermodynamic relation

$$dU = TdS - pdV,$$

we obtain the following expression, after dividing by dV and fixing T :

$$\left(\frac{\partial U}{\partial V} \right)_T = T \left(\frac{\partial S}{\partial V} \right)_T - p = T \left(\frac{\partial p}{\partial T} \right)_V - p.$$

The last equality comes from the following Maxwell relation:

$$\left(\frac{\partial S}{\partial V} \right)_T = \left(\frac{\partial p}{\partial T} \right)_V.$$

From the definition of energy density it follows that

$$U = uV$$

where the energy density of radiation only depends on the temperature, therefore

$$\left(\frac{\partial U}{\partial V}\right)_T = u\left(\frac{\partial V}{\partial V}\right)_T = u\,.$$

Now, the equality

$$\left(\frac{\partial U}{\partial V}\right)_T = T\left(\frac{\partial p}{\partial T}\right)_V - p,$$

after substitution of $\left(\frac{\partial U}{\partial V}\right)_T$ and for the corresponding expressions, can be written as

$$u = \frac{T}{3}\left(\frac{\partial u}{\partial T}\right)_V - \frac{u}{3}.$$

Since the partial derivative $\left(\frac{\partial u}{\partial T}\right)_V$ can be expressed as a relationship between only u and T (if one isolates it on one side of the equality), the partial derivative can be replaced by the ordinary derivative. After separating the differentials the equality becomes

$$\frac{du}{4u} = \frac{dT}{T},$$

which leads immediately to $u = AT^4$, with A as some constant of integration.

Derivation from Planck's Law

The law can be derived by considering a small flat black body surface radiating out into a half-sphere. This derivation uses spherical coordinates, with φ as the zenith angle and θ as the azimuthal angle; and the small flat blackbody surface lies on the xy-plane, where $\varphi = {}^{\pi}/_{2}$.

The intensity of the light emitted from the blackbody surface is given by Planck's law :

$$I(v,T) = \frac{2hv^3}{c^2}\frac{1}{e^{\frac{hv}{kT}} - 1}.$$

where

- $I(v,T)$ is the amount of power per unit surface area per unit solid angle per unit frequency emitted at a frequency v by a black body at temperature T.
- h is Planck's constant
- c is the speed of light, and
- k is Boltzmann's constant.

The quantity $I(v,T)\,A\,dv\,d\Omega$ is the power radiated by a surface of area A through a solid angle $d\Omega$ in the frequency range between v and $v + dv$.

The Stefan–Boltzmann law gives the power emitted per unit area of the emitting body,

$$\frac{P}{A} = \int_0^\infty I(v,T)dv \int d\Omega$$

To derive the Stefan–Boltzmann law, we must integrate Ω over the half-sphere and integrate v from 0 to ∞. Furthermore, because black bodies are *Lambertian* (i.e. they obey Lambert's cosine law), the intensity observed along the sphere will be the actual intensity times the cosine of the zenith angle φ, and in spherical coordinates, $d\Omega = \sin(\varphi)\, d\varphi\, d\theta$.

$$\begin{aligned}\frac{P}{A} &= \int_0^\infty I(v,T)dv \int_0^{2\pi} d\theta \int_0^{\pi/2} \cos\phi \sin\phi d\phi \\ &= \pi \int_0^\infty I(v,T)dv\end{aligned}$$

Then we plug in for I:

$$\frac{P}{A} = \frac{2\pi h}{c^2} \int_0^\infty \frac{v^3}{e^{\frac{hv}{kT}} - 1} dv$$

To do this integral, do a substitution,

$$u = \frac{hv}{kT}$$

$$du = \frac{h}{kT} dv$$

which gives:

$$\frac{P}{A} = \frac{2\pi h}{c^2} \left(\frac{kT}{h}\right)^4 \int_0^\infty \frac{u^3}{e^u - 1} du.$$

The integral on the right is standard and goes by many names: it is a particular case of a Bose-Einstein integral, or the Riemann zeta function, $\zeta(4)$, or the Polylogarithm. The value of the integral is $\frac{\pi^4}{15}$, giving the result that, for a perfect blackbody surface:

$$j^\star = \sigma T^4 \ , \ \sigma = \frac{2\pi^5 k^4}{15 c^2 h^3} = \frac{\pi^2 k^4}{60 \hbar^3 c^2}.$$

Finally, this proof started out only considering a small flat surface. However, any differentiable surface can be approximated by a bunch of small flat surfaces. So long as the geometry of the surface does not cause the blackbody to reabsorb its own radiation, the total energy radiated is just the sum of the energies radiated by each surface; and the total surface area is just the sum of the areas of each surface—so this law holds for all convex blackbodies, too, so long as the surface has the same temperature throughout. The law extends to radiation from non-convex bodies by using the fact that the convex hull of a black body radiates as though it were itself a black body.

References

- C. Carathéodory (1909). "Untersuchungen über die Grundlagen der Thermodynamik". Mathematische Annalen. 67: 355–386. doi:10.1007/BF01450409. A partly reliable translation is to be found at Kestin, J. (1976). The Second Law of Thermodynamics, Dowden, Hutchinson & Ross, Stroudsburg PA
- Thomsen, J.S. (1962). "A restatement of the zeroth law of thermodynamics". Am. J. Phys. 30: 294–296. Bibcode:1962AmJPh..30..294T. doi:10.1119/1.1941991
- Turvey, K. (1990). 'Test of validity of Maxwellian statistics for electrons thermionically emitted from an oxide cathode', European Journal of Physics, 11(1): 51–59. here
- Prati, E. (2010). "The finite quantum grand canonical ensemble and temperature from single-electron statistics for a mesoscopic device". J. Stat. Mech. 1: P01003. Bibcode:2010JSMTE..01..003P. arXiv:1001.2342. doi:10.1088/1742-5468/2010/01/P01003. arxiv.org
- C. Caratheodory (1909). "Untersuchungen über die Grundlagen der Thermodynamik". Mathematische Annalen. 67 (3): 355–386. doi:10.1007/BF01450409
- Swendsen, Robert (March 2006). "Statistical mechanics of colloids and Boltzmann's definition of entropy". American Journal of Physics. 74 (3): 187–190. Bibcode:2006AmJPh..74..187S. doi:10.1119/1.2174962
- Kondepudi, D.K. (1987). "Microscopic aspects implied by the second law". Foundations of Physics. 17: 713–722. Bibcode:1987FoPh...17..713K. doi:10.1007/BF01889544
- Prati, E.; et al. (2010). "Measuring the temperature of a mesoscopic electron system by means of single electron statistics". Applied Physics Letters. 96 (11): 113109. Bibcode:2010ApPhL..96k3109P. arXiv:1002.0037. doi:10.1063/1.3365204. arxiv.org

5

Measurement of Velocity

The instrument used to measure velocity is known as a pitot tube. It is used to measure the speed of an aircraft, boat and the flow of gas in industrial applications. Anemometer, laser Doppler velocimetry and relativistic Doppler effect are some of the topics discussed in the following chapter. The aspects elucidated in this chapter are of vital importance, and provide a better understanding of aerodynamics.

Pitot Tube

A pitot tube is a pressure measurement instrument used to measure fluid flow velocity. The pitot tube was invented by the French engineer Henri Pitot in the early 18th century and was modified to its modern form in the mid-19th century by French scientist Henry Darcy. It is widely used to determine the airspeed of an aircraft, water speed of a boat, and to measure liquid, air and gas flow velocities in industrial applications. The pitot tube is used to measure the local flow velocity at a given point in the flow stream and not the average flow velocity in the pipe or conduit.

Aircraft use pitot tubes to measure airspeed. The example, from an Airbus A380, combines a pitot tube (right) with a static port and an angle-of-attack vane (left). Air-flow is right to left.

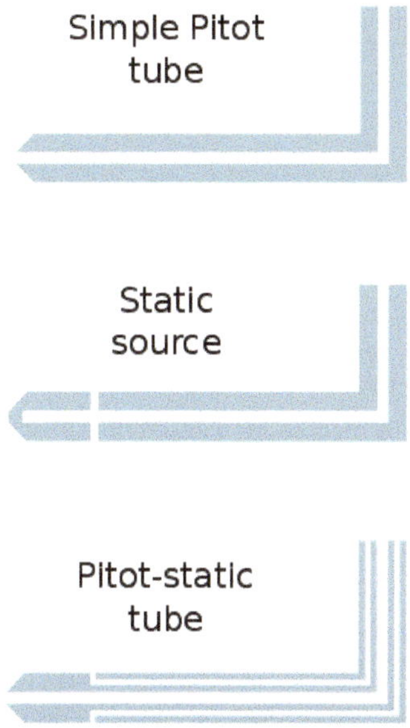

Types of pitot tubes

Pitot tube on Kamov Ka-26 helicopter

Theory of Operation

The basic pitot tube consists of a tube pointing directly into the fluid flow. As this tube contains fluid, a pressure can be measured; the moving fluid is brought to rest (stagnates) as there is no outlet to allow flow to continue. This pressure is the stagnation pressure of the fluid, also known as the total pressure or (particularly in aviation) the pitot pressure.

The measured stagnation pressure cannot itself be used to determine the fluid flow velocity (airspeed in aviation). However, Bernoulli's equation states:

Stagnation pressure = static pressure + dynamic pressure

Which can also be written

$$p_t = p_s + \left(\frac{\rho u^2}{2}\right)$$

Solving that for flow velocity:

$$u = \sqrt{\frac{2(p_t - p_s)}{\rho}}$$

NOTE: The above equation applies only to fluids that can be treated as incompressible. Liquids are treated as incompressible under almost all conditions. Gases under certain conditions can be approximated as incompressible.

where:

- u is flow velocity to be measured in m/s;
- p_t is stagnation or total pressure in pascals;
- p_s is static pressure in pascals;
- and ρ is fluid density in kg / m^3.

The dynamic pressure, then, is the difference between the stagnation pressure and the static pressure. The dynamic pressure is then determined using a diaphragm inside an enclosed container. If the air on one side of the diaphragm is at the static pressure, and the other at the stagnation pressure, then the deflection of the diaphragm is proportional to the dynamic pressure.

In aircraft, The static pressure is generally measured using the static ports on the side of the fuselage. The dynamic pressure measured can be used to determine the indicated airspeed of the aircraft. The diaphragm arrangement described above is typically contained within the airspeed indicator, which converts the dynamic pressure to an airspeed reading by means of mechanical levers.

Instead of separate pitot and static ports, a pitot-static tube (also called a Prandtl tube) may be employed, which has a second tube coaxial with the pitot tube with holes on the sides, outside the direct airflow, to measure the static pressure.

If a liquid column manometer is used to measure the pressure difference $p_t - p_s$, or Δp,

$$\Delta h = \frac{\Delta p}{\rho_l g}$$

where:

- Δh is the height difference of the columns in meters
- ρ_l is the density of the liquid in the manometer
- g is the acceleration of gravity in m / s^2

Therefore,

$$V = \sqrt{\frac{2(\Delta h * (\rho_l g))}{\rho}}$$

Aircraft

A pitot-static system is a system of pressure-sensitive instruments that is most often used in aviation to determine an aircraft's airspeed, Mach number, altitude, and altitude trend. A pitot-static system generally consists of a pitot tube, a static port, and the pitot-static instruments. Errors in pitot-static system readings can be extremely dangerous as the information obtained from the pitot static system, such as airspeed, is potentially safety-critical.

Several commercial airline incidents and accidents have been traced to a failure of the pitot-static system. Examples include Austral Líneas Aéreas Flight 2553, Northwest Airlines Flight 6231, and one of the two X-31s. The French air safety authority BEA said that pitot tube icing was a contributing factor in the crash of Air France Flight 447 into the Atlantic Ocean. In 2008 Air Caraïbes reported two incidents of pitot tube icing malfunctions on its A330s.

Birgenair Flight 301 had a fatal pitot tube failure which investigators suspected was due to insects creating a nest inside the pitot tube; the prime suspect is the Black and yellow mud dauber wasp.

Aeroperú Flight 603 had a pitot-static system failure due to the cleaning crew leaving the static port blocked with tape.

Industry Applications

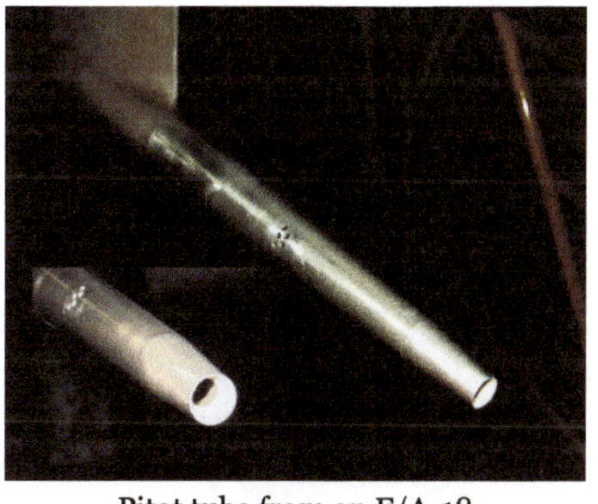

Pitot tube from an F/A-18

Weather instruments at Mount Washington Observatory. Pitot tube static anemometer is on the right.

In industry, the flow velocities being measured are often those flowing in ducts and tubing where measurements by an anemometer would be difficult to obtain. In these kinds of measurements, the most practical instrument to use is the pitot tube. The pitot tube can be inserted through a small hole in the duct with the pitot connected to a U-tube water gauge or some other differential pressure gauge for determining the flow velocity inside the ducted wind tunnel. One use of this technique is to determine the volume of air that is being delivered to a conditioned space.

The fluid flow rate in a duct can then be estimated from:

> Volume flow rate (cubic feet per minute) = duct area (square feet) × flow velocity (feet per minute)
>
> Volume flow rate (cubic meters per second) = duct area (square meters) × flow velocity (meters per second)

In aviation, airspeed is typically measured in knots.

In weather stations with high wind speeds, the pitot tube is modified to create a special type of anemometer called pitot tube static anemometer.

Pneumometric Methods

These methods use a Pitot static tube for assessing the velocity of flow

$$P_0 = p + q$$

$$p_0 - p = \frac{1}{2}\rho v^2$$

$$v = \sqrt{\frac{2\Delta p}{\rho}}$$

$$v = \sqrt{\frac{2\Delta p_i}{\rho}c} \quad \text{..............................1}$$

where p_i is the indicated pressure difference between the Pitot and static pressures and 'c' the tube coefficient.

$$\Delta p_i = \frac{\rho v^2}{2}\frac{1}{c}$$

c is the tube coefficient which depends on the geometry of the tube and position of orifices. c depends on Reynolds number and Mach number.

Measurement in Compressible Flows

The isentropic ratio between total pressure and static pressures is given by

$$\frac{p_0}{p} = \left(1 + \frac{\gamma - 1}{2}M^2\right)^{\gamma/\gamma-1}$$

Expanding using binomial theorem

$$p_0 = p\left(1+\gamma\frac{M^2}{2}+\frac{\gamma}{8}M^4+\frac{\gamma(2-\gamma)}{48}M^6+\ldots\ldots\right)$$

$$= p+\frac{\gamma M^2}{2}p\left[1+\frac{M^2}{4}+\frac{2-\gamma}{24}M^4+\ldots..\right]$$

$$\frac{p}{p}=RT$$

$$\frac{v^2}{a^2}=\frac{v^2}{\gamma RT}=M^2$$

$$\frac{\gamma pM^2}{2}=\frac{\gamma p}{2}\frac{v^2}{\gamma RT}\rightarrow\frac{\rho v^2}{2}$$

$$\frac{p}{RT}=\rho$$

Hence, $p_0 = p+\frac{1}{2}\rho v^2\left[1+\frac{M^2}{4}+\ldots\ldots\right]$ 2

An example:

For $M = 0.2 \simeq 60\ m/s$

The second term becomes 0.01 of $\frac{1}{2}\rho V^2$

$$p_0 - p = \frac{1}{2}\rho v^2(1+\varepsilon)$$

$\varepsilon \rightarrow$ compressibility correction

$$\varepsilon = \frac{M^2}{4}+\frac{(2-\gamma)M^4}{24}+\ldots\ldots\ldots.3$$

which says that in a compressible fluid the difference Δp between total and static pressures exceeds velocity head.

Introducing the tube coefficient c also

$$\frac{p_0}{p}=1+\frac{\gamma}{2}M^2(1+\varepsilon)\times\frac{1}{c}$$

$$\frac{p_0}{p}-1=\frac{\gamma M^2}{2}(1+\varepsilon)\times\frac{1}{c}$$

$$p_0-p=\frac{1}{2}\rho v^2(1+\varepsilon)*\frac{1}{c}$$

$$v=\sqrt{\frac{2\Delta p_i}{\rho(1+\varepsilon)}c} \qquad \ldots\ldots\ldots\ldots\ldots\ldots\ldots\ldots.4$$

Here velocity being measured is expressed as a function of the measured pressure difference, compressibility factor and the tube coefficient.

Measurement of Supersonic Velocities/Mach Number

This is done by measuring inclination angles of shock waves $\beta = \sin^{-1}\dfrac{1}{M}$.

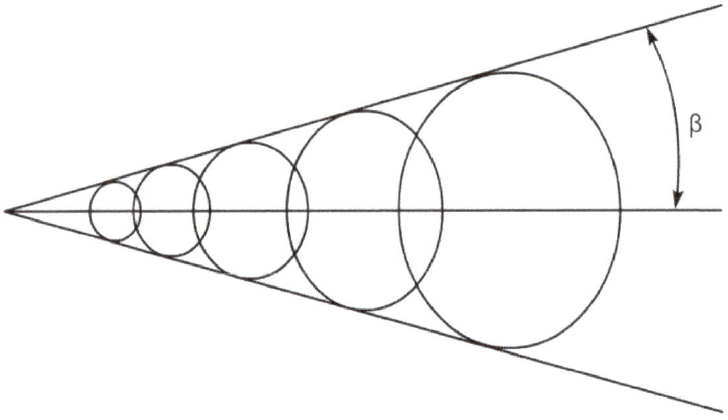

The Mach cone of pressure disturbances

A shock wave of infinitely small intensity lies along the Mach line .But, Mach lines cannot be observed directly. Mach number is best determined by measuring inclination angles of the shock wave appearing at a wedge or a cone shaped obstacle placed with its apex at the test point and axis in the flow direction.

$$\sin^2\beta - \frac{\gamma+1}{2}\frac{\sin\beta\cos\beta}{\cos(\beta-\theta)} = \frac{1}{M^2}$$

In the above equation, θ is the flow deflection angle by the wedge and β is the shock angle.

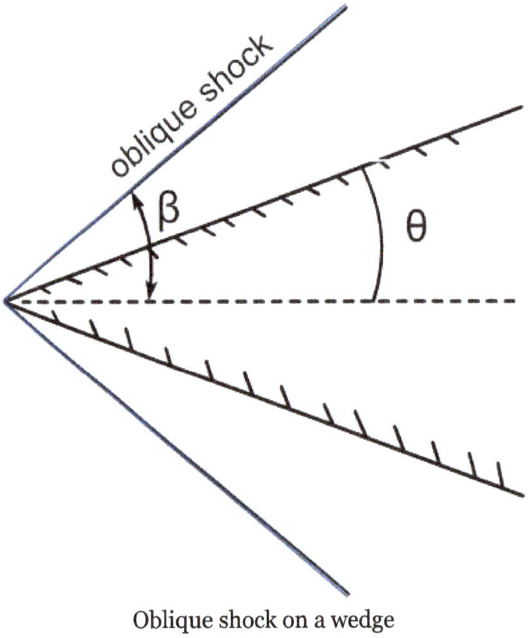

Oblique shock on a wedge

Anemometer

A hemispherical cup anemometer of the type invented in 1846 by John Thomas Romney Robinson

An anemometer is a device used for measuring the speed of wind, and is also a common weather station instrument. The term is derived from the Greek word *anemos*, which means wind, and is used to describe any wind speed measurement instrument used in meteorology. The first known description of an anemometer was given by Leon Battista Alberti in 1450.

History

The anemometer has changed little since its development in the 15th century. Leon Battista Alberti (1404–1472) is said to have invented the first mechanical anemometer around 1450. In following centuries, numerous others, including Robert Hooke (1635–1703), developed their own versions, with some being mistakenly credited as the inventor. In 1846, John Thomas Romney Robinson (1792–1882) improved upon the design by using four hemispherical cups and mechanical wheels. In 1926, Canadian meteorologist John Patterson (January 3, 1872 – February 22, 1956) developed a three-cup anemometer, which was improved by Brevoort and Joiner in 1935. In 1991, Derek Weston added the ability to measure wind direction. In 1994, Andrews Pflitsch developed the sonic anemometer.

Velocity Anemometers

Cup Anemometers

Cup anemometer

A simple type of anemometer was invented in 1845 by Dr. John Thomas Romney Robinson, of Armagh Observatory. It consisted of four hemispherical cups mounted on horizontal arms, which were mounted on a vertical shaft. The air flow past the cups in any horizontal direction turned the shaft at a rate that was proportional to the wind speed. Therefore, counting the turns of the shaft over a set time period produced a value proportional to the average wind speed for a wide range of speeds. On an anemometer with four cups, it is easy to see that since the cups are arranged symmetrically on the end of the arms, the wind always has the hollow of one cup presented to it and is blowing on the back of the cup on the opposite end of the cross.

When Robinson first designed his anemometer, he asserted that the cups moved one-third of the speed of the wind, unaffected by the cup size or arm length. This was apparently confirmed by some early independent experiments, but it was incorrect. Instead, the ratio of the speed of the wind and that of the cups, the anemometer factor, depends on the dimensions of the cups and arms, and may have a value between two and a little over three. Every previous experiment involving an anemometer had to be repeated.

The three-cup anemometer developed by the Canadian John Patterson in 1926 and subsequent cup improvements by Brevoort & Joiner of the USA in 1935 led to a cupwheel design which was linear and had an error of less than 3% up to 60 mph (97 km/h). Patterson found that each cup produced maximum torque when it was at 45° to the wind flow. The three-cup anemometer also had a more constant torque and responded more quickly to gusts than the four-cup anemometer.

The three-cup anemometer was further modified by the Australian Dr Derek Weston in 1991 to measure both wind direction and wind speed. Weston added a tag to one cup, which causes the cupwheel speed to increase and decrease as the tag moves alternately with and against the wind. Wind direction is calculated from these cyclical changes in cupwheel speed, while wind speed is determined from the average cupwheel speed.

Three-cup anemometers are currently used as the industry standard for wind resource assessment studies & practice.

Vane Anemometers

One of the other forms of mechanical velocity anemometer is the *vane anemometer*. It may be described as a windmill or a propeller anemometer. Unlike the Robinson anemometer, whose axis of rotation is vertical, the vane anemometer must have its axis parallel to the direction of the wind and therefore horizontal. Furthermore, since the wind varies in direction and the axis has to follow its changes, a wind vane or some other contrivance to fulfill the same purpose must be employed.

A *vane anemometer* thus combines a propeller and a tail on the same axis to obtain accurate and precise wind speed and direction measurements from the same instrument. The speed of the fan is measured by a rev counter and converted to a windspeed by an electronic chip. Hence, volumetric flowrate may be calculated if the cross-sectional area is known.

In cases where the direction of the air motion is always the same, as in ventilating shafts of mines and buildings, wind vanes known as air meters are employed, and give satisfactory results.

Vane Anemometers

Vane style of anemometer	Helicoid propeller anemometer incorporating a wind vane for orientation	Hand-held low-speed vane anemometer	Hand-held digital anemometer

Hot-wire Anemometers

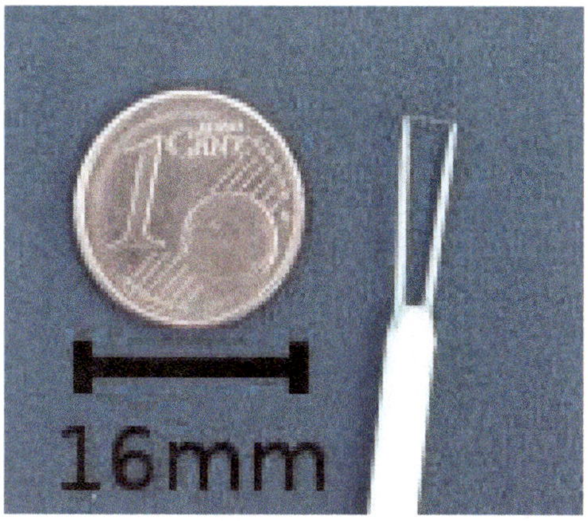

Hot-wire sensor

Hot wire anemometers use a very fine wire (on the order of several micrometres) electrically heated to some temperature above the ambient. Air flowing past the wire cools the wire. As the electrical resistance of most metals is dependent upon the temperature of the metal (tungsten is a popular choice for hot-wires), a relationship can be obtained between the resistance of the wire and the flow speed.

Several ways of implementing this exist, and hot-wire devices can be further classified as CCA (constant current anemometer), CVA (constant voltage anemometer) and CTA (constant-temperature anemometer). The voltage output from these anemometers is thus the result of some sort of circuit within the device trying to maintain the specific variable (current, voltage or temperature) constant, following Ohm's law.

Additionally, PWM (pulse-width modulation) anemometers are also used, wherein the velocity is inferred by the time length of a repeating pulse of current that brings the wire up to a specified resistance and then stops until a threshold "floor" is reached, at which time the pulse is sent again.

Hot-wire anemometers, while extremely delicate, have extremely high frequency-response and fine spatial resolution compared to other measurement methods, and as such are almost universally employed for the detailed study of turbulent flows, or any flow in which rapid velocity fluctuations are of interest.

An industrial version of the fine-wire anemometer is the thermal flow meter, which follows the same concept but uses two pins or stings to monitor the variation in temperature. The strings contain fine wires, but encasing the wires makes them much more durable and capable of accurately measuring air, gas, and emissions flow in pipes, ducts, and stacks. Industrial applications often contain dirt that will damage the classic hot-wire anemometer.

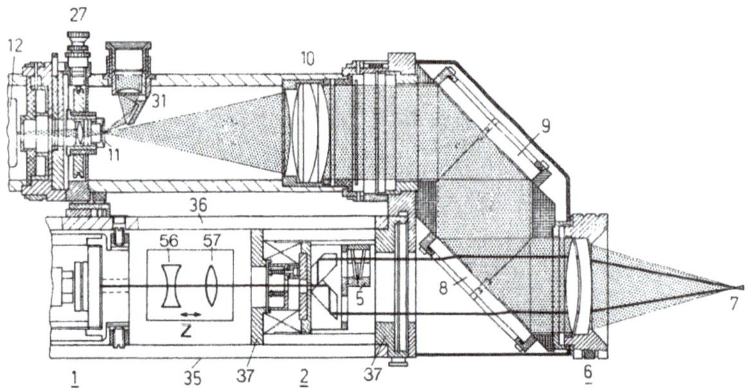

Drawing of a laser anemometer. The laser light is emitted (1) through the front lens (6) of the anemometer and is backscattered off the air molecules (7). The backscattered radiation (dots) re-enters the device and is reflected and directed into a detector (12).

Laser Doppler Anemometers

In laser Doppler velocimetry, laser Doppler anemometers use a beam of light from a laser that is divided into two beams, with one propagated out of the anemometer. Particulates (or deliberately introduced seed material) flowing along with air molecules near where the beam exits reflect, or backscatter, the light back into a detector, where it is measured relative to the original laser beam. When the particles are in great motion, they produce a Doppler shift for measuring wind speed in the laser light, which is used to calculate the speed of the particles, and therefore the air around the anemometer.

2D ultrasonic anemometer with 3 paths

Ultrasonic Anemometers

3D ultrasonic anemometer

Ultrasonic anemometers, first developed in the 1950s, use ultrasonic sound waves to measure wind velocity. They measure wind speed based on the time of flight of sonic pulses between pairs of transducers. Measurements from pairs of transducers can be combined to yield a measurement of velocity in 1-, 2-, or 3-dimensional flow. The spatial resolution is given by the path length between transducers, which is typically 10 to 20 cm. Ultrasonic anemometers can take measurements with very fine temporal resolution, 20 Hz or better, which makes them well suited for turbulence measurements. The lack of moving parts makes them appropriate for long-term use in exposed automated weather stations and weather buoys where the accuracy and reliability of traditional cup-and-vane anemometers are adversely affected by salty air or dust. Their main disadvantage is the distortion of the flow itself by the structure supporting the transducers, which requires a correction based upon wind tunnel measurements to minimize the effect. An international standard for this process, ISO 16622 *Meteorology—Ultrasonic anemometers/thermometers—Acceptance test methods for mean wind measurements* is in general circulation. Another disadvantage is lower accuracy due to precipitation, where rain drops may vary the speed of sound.

Since the speed of sound varies with temperature, and is virtually stable with pressure change, ultrasonic anemometers are also used as thermometers.

Two-dimensional (wind speed and wind direction) sonic anemometers are used in applications such as weather stations, ship navigation, wind turbines, aviation and weather buoys. Three-dimensional sonic anemometers are widely used to measure gas emissions and ecosystem fluxes using the eddy covariance method when used with fast-response infrared gas analyzers or laser-based analyzers.

Two-dimensional wind sensors are of two types:

- Two ultrasounds paths: These sensors have 4 arms. This design is more common, due to its simplicity. The disadvantage of this type of sensor is that when the wind comes in the direction of an ultrasound path, the arms disturb the airflow, reducing the accuracy of the resulting measurement.

- Three ultrasounds paths: These sensors have 3 arms. They give one path redundancy of the measurement which improves the sensor accuracy and reduces aerodynamic turbulence.

Acoustic Resonance Anemometers

Acoustic resonance anemometer

Acoustic resonance anemometers are a more recent variant of sonic anemometer. The technology was invented by Dr Savvas Kapartis and patented (Acu-Res®) by FT Technologies in 2000. Whereas conventional sonic anemometers rely on time of flight measurement, acoustic resonance sensors use resonating acoustic (ultrasonic) waves within a small purpose-built cavity in order to perform their measurement.

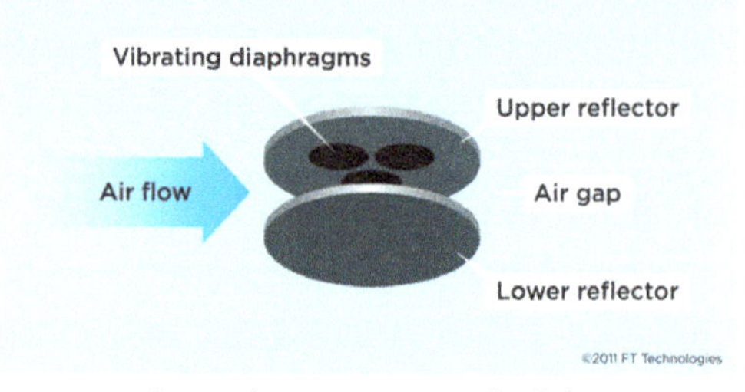

Acoustic resonance principle

Built into the cavity is an array of ultrasonic transducers, which are used to create the separate standing-wave patterns at ultrasonic frequencies. As wind passes through the cavity, a change in the wave's property occurs (phase shift). By measuring the amount of phase shift in the received signals by each transducer, and then by mathematically processing the data, the sensor is able to provide an accurate horizontal measurement of wind speed and direction.

Acoustic resonance technology enables measurement within a small cavity, the sensors therefore tend to be typically smaller in size than other ultrasonic sensors. The small size of acoustic resonance anemometers makes them physically strong and easy to heat and therefore resistant to icing. This combination of features means that they achieve high levels of data availability and are well suited to wind turbine control and to other uses that require small robust sensors such as battlefield meteorology. One issue with this sensor type is measurement accuracy when compared to a calibrated mechanical sensor. For many end uses, this weakness is compensated for by the sensor's longevity and the fact that it does not require re-calibrating once installed.

Ping-pong Ball Anemometers

A common anemometer for basic use is constructed from a ping-pong ball attached to a string. When the wind blows horizontally, it presses on and moves the ball; because ping-pong balls are very lightweight, they move easily in light winds. Measuring the angle between the string-ball apparatus and the vertical gives an estimate of the wind speed.

This type of anemometer is mostly used for middle-school level instruction which most students make themselves, but a similar device was also flown on Phoenix Mars Lander.

Pressure Anemometers

Britannia Yacht Club clubhouse tour, burgee, and wind gauge on roof

The first designs of anemometers which measure the pressure were divided into plate and tube classes.

Plate Anemometers

These are the first modern anemometers. They consist of a flat plate suspended from the top so that the wind deflects the plate. In 1450, the Italian art architect Leon Battista Alberti invented the first mechanical anemometer; in 1664 it was re-invented by Robert Hooke (who is often mistakenly considered the inventor of the first anemometer). Later versions of this form consisted of a flat plate, either square or circular, which is kept normal to the wind by a wind vane. The pressure of the wind on its face is balanced by a spring. The compression of the spring determines the actual force which the wind is exerting on the plate, and this is either read off on a suitable gauge, or on a recorder. Instruments of this kind do not respond to light winds, are inaccurate for high wind readings, and are slow at responding to variable winds. Plate anemometers have been used to trigger high wind alarms on bridges.

Tube Anemometers

Tube anemometer invented by William Henry Dines. The movable part (right) is mounted on the fixed part (left).

Instruments at Mount Washington Observatory. The pitot tube static anemometer is on the right.

The pointed head is the pitot port. The small holes are connected to the static port.

James Lind's anemometer of 1775 consisted of a glass U tube containing a liquid manometer (pressure gauge), with one end bent in a horizontal direction to face the wind and the other vertical end remains parallel to the wind flow. Though the Lind was not the first it was the most practical and best known anemometer of this type. If the wind blows into the mouth of a tube it causes an increase of pressure on one side of the manometer. The wind over the open end of a vertical tube causes little change in pressure on the other side of the manometer. The resulting elevation difference in the two legs of the U tube is an indication of the wind speed. However, an accurate measurement requires that the wind speed be directly into the open end of the tube; small departures from the true direction of the wind causes large variations in the reading.

The successful metal pressure tube anemometer of William Henry Dines in 1892 utilized the same pressure difference between the open mouth of a straight tube facing the wind and a ring of small holes in a vertical tube which is closed at the upper end. Both are mounted at the same height. The pressure differences on which the action depends are very small, and special means are required to register them. The recorder consists of a float in a sealed chamber partially filled with water. The pipe from the straight tube is connected to the top of the sealed chamber and the pipe from the small tubes is directed into the bottom inside the float. Since the pressure difference determines the vertical position of the float this is a measure of the wind speed.

The great advantage of the tube anemometer lies in the fact that the exposed part can be mounted on a high pole, and requires no oiling or attention for years; and the registering part can be placed in any convenient position. Two connecting tubes are required. It might appear at first sight as though one connection would serve, but the differences in pressure on which these instruments depend are so minute, that the pressure of the air in the room where the recording part is placed has to be considered. Thus if the instrument depends on the pressure or suction effect alone, and this pressure or suction is measured against the air pressure in an ordinary room, in which the doors and windows are carefully closed and a newspaper is then burnt up the chimney, an effect may be produced equal to a wind of 10 mi/h (16 km/h); and the opening of a window in rough weather, or the opening of a door, may entirely alter the registration.

While the Dines anemometer had an error of only 1% at 10 mph (16 km/h) it did not respond very well to low winds due to the poor response of the flat plate vane required to turn the head into the wind. In 1918 an aerodynamic vane with eight times the torque of the flat plate overcame this problem.

Pitot Tube Static Anemometers

Modern tube anemometers use the same principle as in the Dines anemometer but using a different design. The implementation uses a pitot-static tube which is a pitot tube with two ports, pitot and static, that is normally used in measuring the airspeed of aircraft. The pitot port measures the dynamic pressure of the open mouth of a tube with pointed head facing wind, and the static port measures the static pressure from small holes along the side on that tube. The pitot tube is connected to a tail so that it always makes the tube's head to face the wind. Additionally, the tube is heated to prevent rime ice formation on the tube. There are two lines from the tube down to the devices to measure the difference in pressure of the two lines. The measurement devices can be manometers, pressure transducers, or analog chart recorders.

Effect of Density on Measurements

In the tube anemometer the dynamic pressure is actually being measured, although the scale is usually graduated as a velocity scale. If the actual air density differs from the calibration value, due to differing temperature, elevation or barometric pressure, a correction is required to obtain the actual wind speed. Approximately 1.5% (1.6% above 6,000 feet) should be added to the velocity recorded by a tube anemometer for each 1000 ft (5% for each kilometer) above sea-level.

Effect of Icing

At airports, it is essential to have accurate wind data under all conditions, including freezing precipitation. Anemometry is also required in monitoring and controlling the operation of wind turbines, which in cold environments are prone to in-cloud icing. Icing alters the aerodynamics of an anemometer and may entirely block it from operating. Therefore, anemometers used in these applications must be internally heated. Both cup anemometers and sonic anemometers are presently available with heated versions.

Instrument Location

In order for wind speeds to be comparable from location to location, the effect of the terrain needs to be considered, especially in regard to height. Other considerations are the presence of trees, and both natural canyons and artificial canyons (urban buildings). The standard anemometer height in open rural terrain is 10 meters.

Hot Wire Anemometer (HWA)

The hot wire anemometer is used to measure fluid velocities by measuring heat loss by convection from a very fine wire which is exposed to the fluid stream. The wire is electrically heated by passing an electrical current through it. When the heated wire is cooled by a fluid stream its electrical resistance decreases, because the resistance of metal wire varies linearly with its temperature.

Basic work on HWA

One of the basic papers on HWA is by King L.V.

Heat transfer from a heated cylinder in an airstream.

Nusselts number $= Nu = \dfrac{Heat\ loss\ /\ unit\ length}{\pi K_f\left(T_w - T_f\right)}$

T_W = Temperature of wire

T_f = Temperature of fluid

K_f = Thermal conductivity of fluid

Assumptions

1) Cylinder is infinitely long

2) No heat loss due to radiation (valid as T_W 200°C)

According to King's Law,

$Nu = A + BR_e^{1/2}$ 5

where A and B are constants

$R_e = \dfrac{\bar{u}d}{\gamma_f}$

d is diameter of wire.

γf → kinematic viscosity

Advantages of HWA

a) Compared to pneumometric methods, HWA has fast response. Manometers have transmission lag to a more or less extent depending on the measurement system employed. Instantaneous measurement of velocity is important for time dependent fluid phenomena such as turbulence.

b) In pneumometric methods, sensitivity decreases as the velocity decreases. In the case of HWA, sensitivity is more at lower velocities.

c) Small probes can be made.

Probes used in HWA

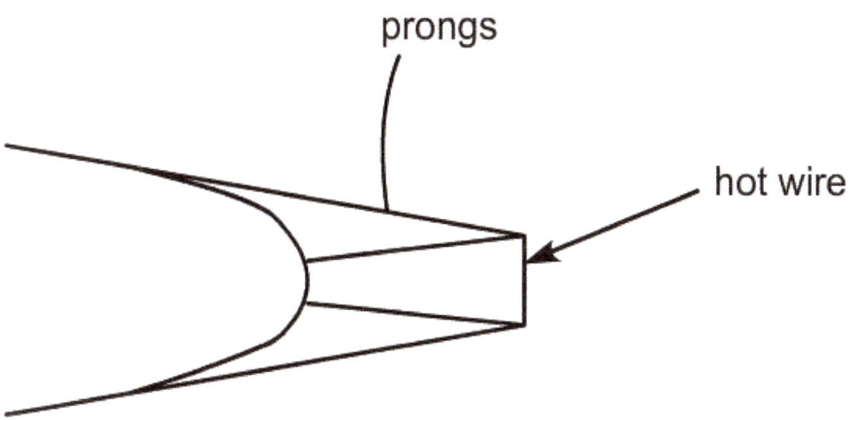

Hot wire probe

The material of the hot wire is Tungsten, Platinum or Platinum Rhodium

Response to Fluctuation

Response to fluctuation depends on wire diameter. Smaller the diameter faster is the response. Diameter of the hot wire is usually between 2.5 and 10μ [say~ 5μ]

To minimise heat loss to prongs

Length of wire $l_w >> d_w$

Usual length $l \simeq 1\ to\ 1.2mm$

$$\frac{l}{d} << 240$$

Two Modes of HWA

a) Constant current anemometer

In this case, the current through the wire is kept constant. When the flow takes place over the hot wire, changes in temperature and resistance occur. The resultant change in voltage is calibrated against velocity.

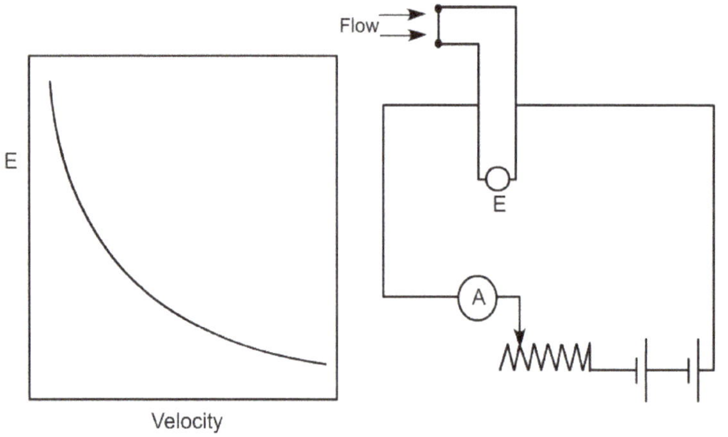

Constant current anemometer and a typical calibration curve

I. Due to thermal inertia the response is limited to about 1000Hz.

II. The current is usually set high to heat the wire considerably above fluid temperature. If the flow drops or velocity becomes zero the wire might damaged due to burn out as there is no convection transfer.

b) Constant temperature anemometer

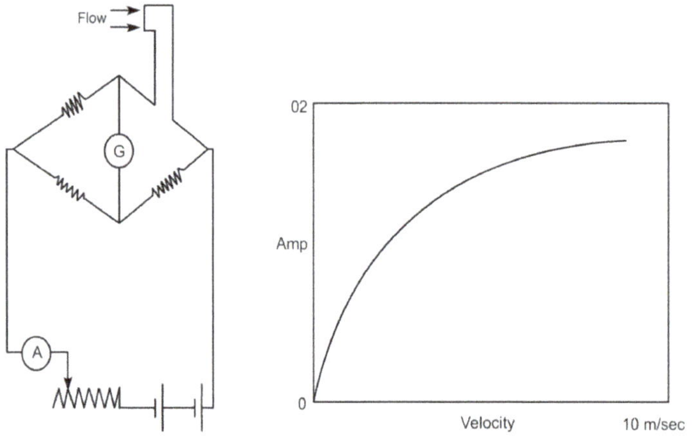

Constant temperature anemometer and a typical calibration curve

A change in velocity of the flow over the hot wire causes temperature and resistance to change. This imbalances the bridge. By adjusting the resistance of the adjacent arm the bridge is rebalanced. In other words the wire temperature is restored. In constant temperature (resistance) method, the velocity is determined in terms of current needed to maintain a constant temperature of the wire.

Elementary Physics of Laser Action

Before the next method of velocity measurement, the Laser Doppler Anemometer, is explained it is useful to have an understanding of the basic laser action itself.

Consider a collection of molecules in a gas. Pick one of the molecules and examine it. The molecule is moving about in space – it has translational energy – it is rotating about its principal axes. So, it has rotational kinetic energy. The atoms that make up the molecule may be vibrating back and forth from some equilibrium position – it has vibrational kinetic and potential energy. Electrons move about the nuclei of the molecule hence, it has kinetic and potential energy of electronic origin. One amazing quality of these various forms of molecular energy is that they can not be any arbitrary value. Molecule at any instant has to occupy one of the very specific set of energy levels.

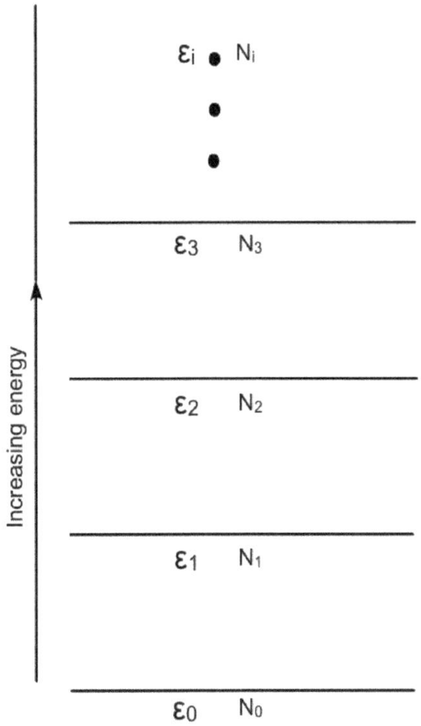

Population distribution at different energy levels

The numbers N_o, N_1, N_2, N_3 .... N_i is called the population distribution over the energy levels ε_o ...... ε_i of the gas. Consider the vibrational energy of a molecular gas. If the gas is in thermodynamic equilibrium, the population distribution will exponentially decrease with increasing Ni that is it will follow a Boltsman distribution.

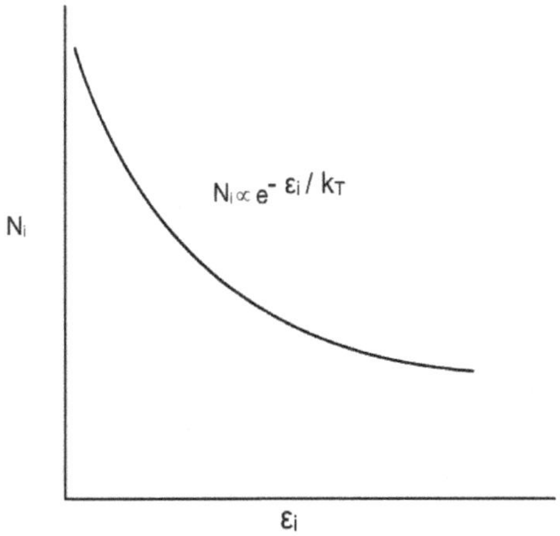

Boltsman distribution of energy

The nature of this population distribution is of vital importance for laser action. A major characteristic of the equilibrium distribution is that $N_{i+1} < N_i$. However, if the gas is disturbed at any instant say by a temperature change and by an electric discharge then the population distribution and in some cases if it is possible that $N_{i+1} > N_i$. This situation in which the number of molecules in a higher energy level is greater than the number in a lower level is called population inversion. The population inversion is the essence of laser action. The attainment of population inversion by means of rapid cooling of the gas is the essence of a G.D laser.

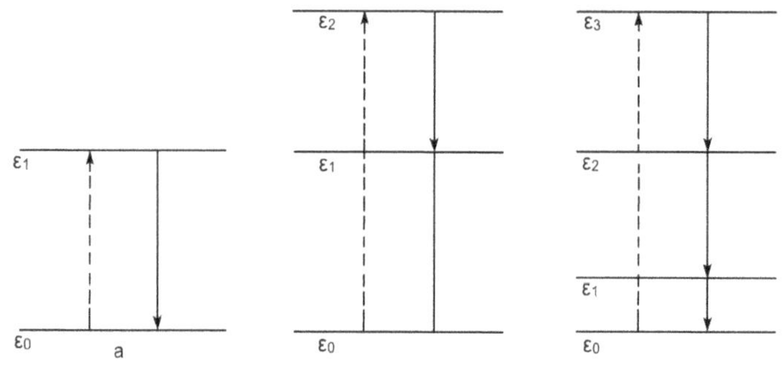

Excited energy levels

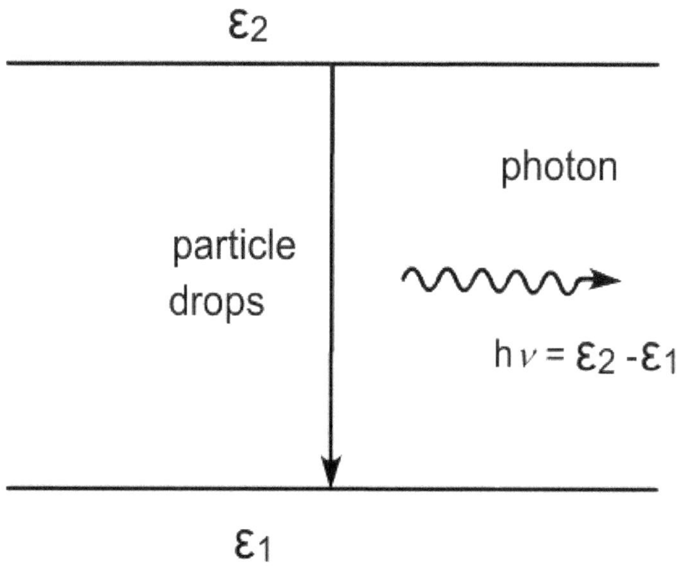

Representation of spontaneous emission

Let us consider one electron of the many in an atom E_O is the ground state of the electron the one in which it normally resides. E_1 is the state (excited state) into which the electron can be driven by supplying it with energy. This energy can come in the form of a photon. The frequency v of light which can excite the electron is given by $E_1 - E_O = hv$.

If a light wave of frequency v passes through a material containing these it will excite them and be absorbed. Higher energy states are unstable and after a short time the electron will spontaneously revert to the ground state emitting a photon v in the process. The average time an atom or molecule spends in the excited state is spontaneous emission lifetime. The net result is that light travelling in one direction will be absorbed and re-emitted in all directions. Instead of leaving the excited atom to decay naturally it is irradiated with another photon.

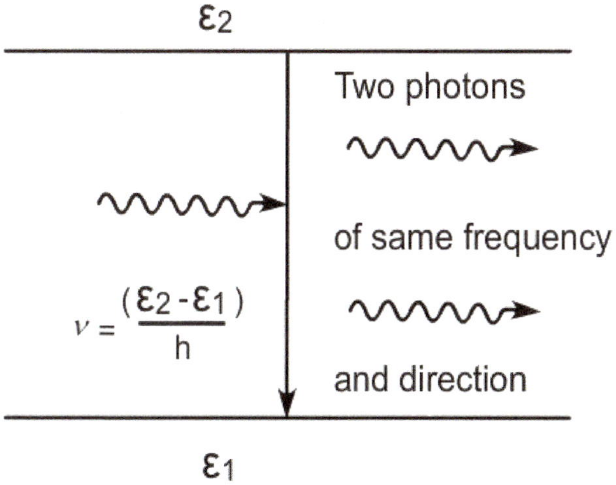

Representation of stimulated emission

Another possibility can however occur. A photon of frequency v can stimulate an already excited atom to emit a second photon travelling in the same direction as the first, whilst the atom reverts to the ground state. This is the stimulated emission upon which a laser relies. If more electrons are in the excited state than the ground state, the net result is amplification. So, the problem of constructing a laser is then essentially one of creating population inversion. If there are only two levels possible for the electron, an enormous amount of energy would be required to excite the majority of the electrons into the upper state.

Therefore, one has to cheat by using a three level system. The electrons are excited by a lamp, electric discharge or other means from ε_0 *to* ε_2 There are no electrons in ε level. So, a population inversion exists between ε_2 *and* ε_1. If a photon of frequency $\frac{(\varepsilon_2 - \varepsilon_1)}{h}$ passes through the medium it will produce stimulated emission.

This is stimulated emission. If max atoms or molecules are in the excited state, the net result is amplification. Three basic requirements for any system to operate as Laser are

- A material must be available with a suitable energy level system.
- A means of producing population inversion must be devised.
- A sufficiently large number of atoms in the excited level must be available if stimulated emission is to exceed spontaneous emission.

Characteristics of Laser Light

1 High intensity

2 Directionality

3 Coherence

4 Monochromaticity

Laser Doppler Velocimetry

Laser Doppler velocimetry (LDV), also known as laser Doppler anemometry (LDA), is the technique of using the Doppler shift in a laser beam to measure the velocity in transparent or semi-transparent fluid flows, or the linear or vibratory motion of opaque, reflecting, surfaces. The measurement with LDA is absolute, linear with velocity and requires no pre-calibration.

LDA facility operating at Laboratory of Gas Technology (Poznań University of Technology).

Technology Origin

With the development of the helium–neon laser (He-Ne) at the Bell Telephone Laboratories in 1962, the optics community had available a source of continuous wave electromagnetic radiation highly concentrated at a wavelength of 632.8 nanometers (nm), in the red portion of the visible spectrum. It was soon shown fluid flow measurement could be made from the Doppler effect on a He-Ne beam scattered by very small polystyrene spheres entrained in the fluid.

At the Research Laboratories of Brown Engineering Company (later Teledyne Brown Engineering), this phenomenon was used in developing the first laser Doppler flowmeter using heterodyne signal processing.

The instrument was soon called the laser Doppler velocimeter (LDV) and the technique laser Doppler velocimetry, also abbreviated LDV. Another application name is laser Doppler anemometry (LDA). Early LDV applications ranged from measuring and mapping the exhaust from rocket engines with speeds up to 1000 m/s to determining flow in a near-surface blood artery. A variety of similar instruments were developed for solid-surface monitoring, with applications ranging from measuring product speeds in production lines of paper and steel mills, to measuring vibration frequency and amplitude of surfaces.

Operating Principles

In its simplest and most presently used form, LDV crosses two beams of collimated, monochromatic, and coherent laser light in the flow of the fluid being measured. The two beams are usually obtained by splitting a single beam, thus ensuring coherence between the two. Lasers with wavelengths in the visible spectrum (390–750 nm) are commonly used; these are typically He-Ne, Argon ion, or laser diode, allowing the beam path to be observed. A transmitting optics focuses the beams to intersect at their waists (the focal point of a laser beam), where they interfere and generate a set of straight fringes. As particles (either naturally occurring or induced) entrained in the fluid pass through the fringes, they reflect light that is then collected by a receiving optics and focused on a photodetector (typically an avalanche photodiode).

The reflected light fluctuates in intensity, the frequency of which is equivalent to the Doppler shift between the incident and scattered light, and is thus proportional to the component of particle velocity which lies in the plane of two laser beams. If the sensor is aligned to the flow such that the fringes are perpendicular to the flow direction, the electrical signal from the photodetector will then be proportional to the full particle velocity. By combining three devices (e.g.; He-Ne, Argon ion, and laser diode) with different wavelengths, all three flow velocity components can be simultaneously measured.

Another form of LDV, particularly used in early device developments, has a completely different approach akin to an interferometer. The sensor also splits the laser beam into two parts; one (the measurement beam) is focused into the flow and the second (the reference beam) passes outside the flow. A receiving optics provides a path that intersects the measurement beam, forming a small volume. Particles passing through this volume will scatter light from the measurement beam with a Doppler shift; a portion of this light is collected by the receiving optics and transferred to the photodetector. The reference beam is also sent to the photodetector where optical heterodyne detection produces an electrical signal proportional to the Doppler shift, by which the particle velocity component perpendicular to the plane of the beams can be determined.

Similar arrangements using optical heterodyning are also used in laser Doppler sensors for measuring the linear velocity of solids and for measuring vibrations of surfaces; the latter sensor is usually called a laser Doppler vibrometer, also abbreviated LDV.

Applications

In the decades since the LDV was first introduced, there has been a wide variety of laser Doppler sensors developed and applied.

Flow Research

Laser Doppler velocimetry is often chosen over other forms of flow measurement because the equipment can be outside of the flow being measured and therefore has no effect on the flow. Some typical applications include the following:

- Wind tunnel velocity experiments for testing aerodynamics of aircraft, missiles, cars, trucks, trains, and buildings and other structures

- Velocity measurements in water flows (research in general hydrodynamics, ship hull design, rotating machinery, pipe flows, channel flow, etc.).
- Fuel injection and spray research where there is a need to measure velocities inside engines or through nozzles
- Environmental research (combustion research, wave dynamics, coastal engineering, tidal modeling, river hydrology, etc.).

One disadvantage has been that LDV sensors are range-dependent; they have to be calibrated minutely and the distances where they measure has to be precisely defined. This distance restriction has recently been at least partially overcome with a new sensor that is range independent.

Automation

Laser Doppler velocimetry can be useful in automation, which includes the flow examples above. It can also be used to measure the speed of solid objects, like conveyor belts. This can be useful in situations where attaching a rotary encoder (or a different mechanical speed measurement device) to the conveyor belt is impossible or impractical.

Medical Applications

Laser Doppler velocimetry is used in hemodynamics research as a technique to partially quantify blood flow in human tissues such as skin. Within the clinical environment, the technology is often referred to as laser Doppler flowmetry (LDF). The beam from a low-power laser (usually a laser diode) penetrates the skin sufficiently to be scattered with a Doppler shift by the red blood cells and return to be concentrated on a detector. These measurements are useful to monitor the effect of exercise, drug treatments, environmental, or physical manipulations on targeted micro-sized vascular areas.

The laser Doppler vibrometer is being used in clinical otology for the measurement of tympanic membrane (eardrum), malleus (hammer), and prosthesis head displacement in response to sound inputs of 80- to 100-dB sound-pressure level. It also has potential use in the operating room to perform measurements of prosthesis and stapes (stirrup) displacement.

Navigation

The Autonomous Landing Hazard Avoidance Technology used in NASA's Project Morpheus lunar lander to automatically find a safe landing place contains a lidar Doppler velocimeter that measures the vehicle's altitude and velocity. The AGM-129 ACM cruise missile uses laser doppler velocimeter for precise terminal guidance.

Laser Doppler Anemometry is the measurement of fluid velocities by detecting frequency shift of laser light that has been scattered by small particles moving with the fluid. Determining velocity by observing the frequency shift of scattered light was first discussed in a paper by Cummins, Knable and Yeh titled "Observation of Diffusion broadening of Rayleigh Scattered light". Their study was on the Brownian motion of an aqueous suspension of micron sized particles by observing the spectrum of scattered light. They observed the net shift in the frequency of light which they attributed to the small convection currents that generated mean velocity in their cell. Shortly thereafter, Yeh and Cummins carried out experiments intended expressly to demonstrate the measurement of fluid velocities.

Advantages of LDA for Velocity Measurements

(i) The major advantage is that the method is non-intrusive. It could be used in flows that are hostile to material probes or that would be altered by material probe.

(ii) No calibration needed.

(iii) Many components can be measured.

(iv) Good spatial resolution.

(v) Capable of tracking very high frequency fluctuations [with fast electronics].

(vi) Impressive accuracies achievable. Reported up to 0.1% absolute accuracy

Versatility of LDA is illustrated by the types of flows such as the following in which it has been used successfully

- Supersonic flows
- Recirculating flows
- Natural free convection
- Flow in IC engines
- Chemically reacting flows
- Two phase flows
- Atmospheric turbulence
- High temperature plasmas

Limitations

(a) Weak intensities of light scattered by small particles resulted in noisy signals.

(b) Random locations of the scattering particles in the fluid created new problems of data analysis. Data arrive randomly and infrequently.

(c) Optical arrangement is sensitive to vibrations.

Doppler Shift: Simplistic Physical Explanation

Consider a wave motion being received by a person at P.

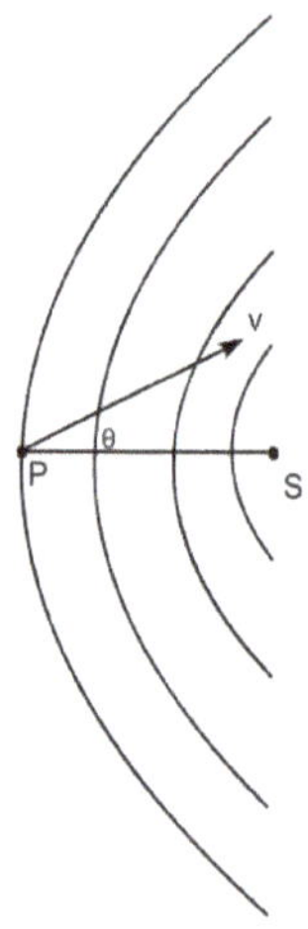

Propagation of waves from a source

$$f = \frac{c}{\lambda}$$

In unit time the distance travelled by observer P in the direction of source S is $v\cos\theta$ where, θ is the angle, the velocity vector makes with the direction of wave motion.

So, in unit time $\dfrac{v\cos\theta}{\lambda}$ would be intercepted by P than it had remained stationary. So, change in frequency apparent to the observer is

$$\Delta f = \frac{v\cos\theta}{\lambda} = \frac{f_s}{c} v\cos\theta \quad \text{.............................6}$$

$$f = \frac{c}{\lambda}$$

Difference in frequency from the original is caused by the movement of P.

Classical Approach to Doppler Shift Equation

An electromagnetic radiation scattered by a moving object suffers a change in frequency in proportion to the velocity of the scattering object.

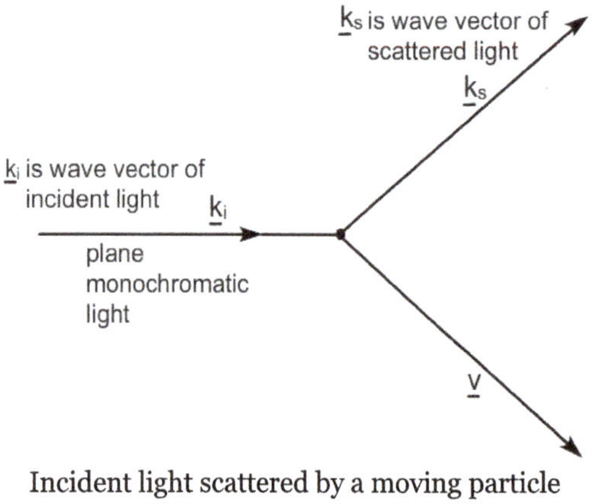

Incident light scattered by a moving particle

Assume that a plane monochromatic wave is incident on a particle moving with velocity $\underline{v}$ such that $\underline{v}$ is much less than c the speed of light. For a stationary particle, the number of wave fronts incident upon it per unit time.

$$f_i = c/\lambda_i \quad \text{...} 7$$

The number of wave fronts incident upon a moving particle per unit time.

$$f_p = \frac{c - \underline{v}.\underline{i}}{\lambda_i} \quad \text{.......................................} 8$$

f_i = frequency of incident light

λ_i = wavelength of incident light

$c-\underline{v}.\underline{i}$ = velocity difference between the particles and the illuminating wave

$\underline{i}$ = unit vector parallel to k_i the wave vector of incident light

The wavelength λ_p apparent to the particle is $\lambda_p = \dfrac{c}{f_p}$

Substitute from (8)

$$\lambda_p = \frac{\lambda_i c}{c - \underline{v}.\underline{i}} \quad \text{..} 9$$

For a stationary observer viewing along ks the apparent scattered wavelength is

$$\lambda_S = \frac{c - \underline{v}.\underline{s}}{f_p} \quad \text{..} 10$$

$\underline{v}.\underline{s}$ represents the amount of particle velocity along the direction $\underline{s}$.

$\underline{s}$ is unit vector parallel to $\underline{k}s$ which is wave vector of scattered light. Hence, frequency of scattered radiation.

$$f_s = \frac{c}{\lambda_s} = \frac{c^* f_p}{c - \underline{v}.\underline{s}} \quad \text{(Equation 7.10 is substituted to get this equation)}$$

$$\therefore f_s = \frac{c(c - \underline{v}.\underline{i})}{\lambda_i (c - \underline{v}.\underline{s})} \rightarrow f_p \text{ is substituted from equation (8)..............} 11$$

The change in frequency or the Doppler shift is

$$f_D = f_S - f_i \qquad ie. \qquad \frac{c(c - \underline{v}.\underline{i})}{\lambda(c - \underline{v}.\underline{s})} - \frac{c}{\lambda_i}$$

$$= \frac{c}{\lambda_i}\left(\frac{c-\underline{v}.\underline{i}}{c-\underline{v}.\underline{s}}-1\right) \quad \text{.......................12}$$

$$\text{For } |v| << c \rightarrow \frac{c}{\lambda_i}\left(\frac{\not{c}-\underline{v}.\underline{i}-\not{c}+\underline{v}.\underline{s}}{c-\underline{v}.\underline{s}}\right) = \frac{c}{\lambda_i}\left(\frac{\underline{v}.\underline{s}-v.\underline{i}}{c-\underline{v}.\underline{s}}\right)$$

$$f_D = \frac{\underline{v}.(\underline{s}-\underline{i})}{\lambda_i} \quad \text{....................................13}$$

Intensity of the Scattered Light

The intensity of light scattered by particles depends upon:

I. The total incident intensity illuminating the particle.

II. The scattering angle.

III. The normalized particle diameter where, is the wavelength of the incident radiation.

IV. The ratio m of the refractive indices of the scatterer and the surrounding medium.

In anemometer applications the diameter of particles is usually comparable to or greater then the wavelength of incident radiation. Foreman and co-workers found that, the scattering is much greater for small angles in the forward direction. For scattering angles less than 10° the intensity is several hundred times greater than for scattering through 180°.

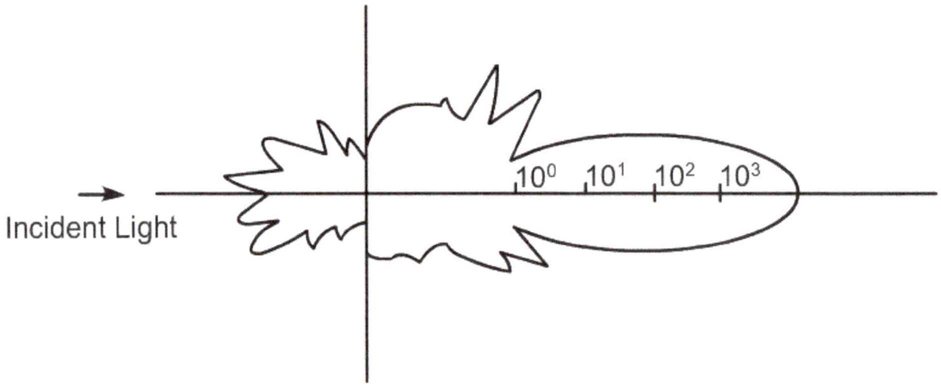

Typical example of spatial intensity distribution of scattered light

Two Modes of LDA

Forward and Backward Scattering Modes of LDA

In the forward scattering mode the scattered light is collected by the photo detector at small degrees in the forward direction. As there is higher intensity in this direction it is possible to build the LDA system using lower power laser. The drawback with this arrangement is that both sides of the test set up will be engaged with the equipments.In the case of backward scattering mode,the photo detector is also kept at the same side of the laser. As the scattered intensity is lower in this direction, the power of the laser source has to be more

Basic LDA Arrangement

To understand the basic arrangement of LDA, refer to figure given below.

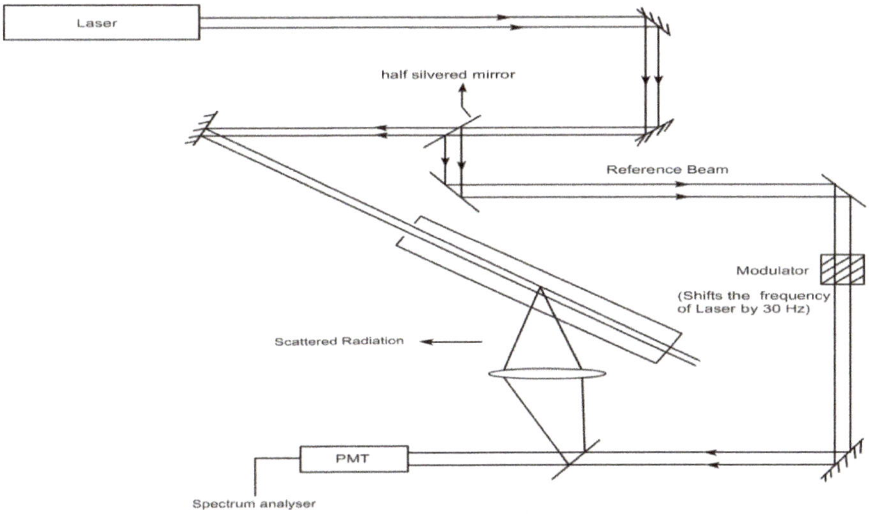

Basic arrangement of LDA

Light from a helium Neon laser is used at a wavelength of 632.8nm which is red light. The light is scattered polystyrene spheres approximately 0.5 μm in diameter which are suspended in water at a concentration of 1 part in 30,000 by volume.

The beam is split by a half silvered mirror. One beam passes through the flow tube. This laser light is scattered by the fluid and collected through a part in the flow tube. This signal beam is focused upon the photo cathode of a photomultiplier. The other beam is passed through a single side band modulator which shifts the frequency of the laser light by 30MHz.

The reference beam is then caused to fall upon the photomultiplier and mix with the signal beam. For zero velocity a frequency shift of 30MHz will be observed. The Doppler frequency shift is added to this 30MHz. Yeh and Cummins used velocimeter to measure velocity profile of laminar flow through the flow tube.

Single Beam Laser Anemometer

The first extensive paper on a Laser anemometer was written by Foreman, George, Lewis, Thorton, Felton and Watson IEE Jl. These researchers used what is now termed a single-beam anemometer. The schematic of the single beam laser anemometer is given in figure below.

Once beam from the laser is focused upon the fluid whose velocity is to be measured, part of the beam is scattered and part is transmitted to produce the reference beam. The signal beam is collimated by a lens L_1, passes through an aperture A which defines the scattering angle and is reflected by a mirror M_1 onto a photo multiplier tube. The reference beam is collimated by a lens, L_2 and made to overlap the signal beam by means of the mirror M_2 and a beam splitter.

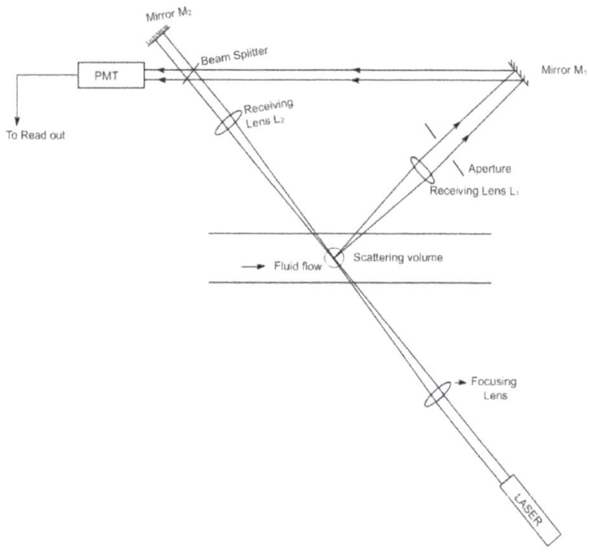

Single beam Laser Doppler Anemometer

Foreman and co-workers after extensive studies on the effect of angle on scattering intensity concluded that scattering at small angles gives much greater intensity. Scattering at less than 10° was found to be several hundred times greater than scattering through 180°. Accordingly their system and all subsequent systems used light which was scattered through only small angles.

Dual Beam Laser Anemometer

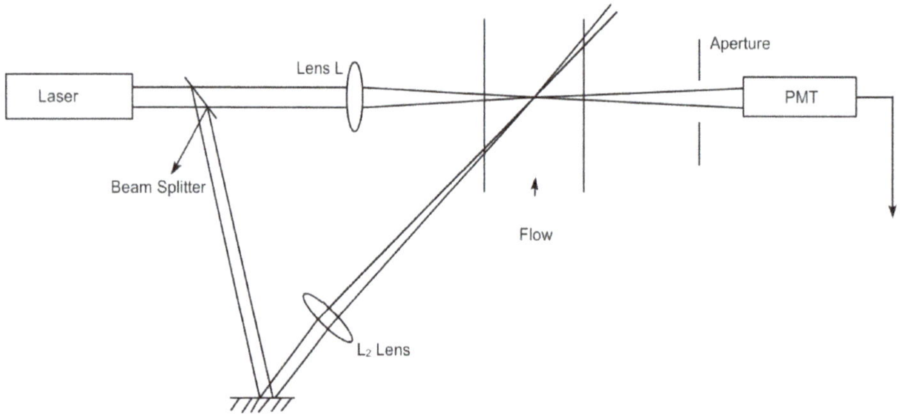

Schematic of a dual beam anemometer

The schematic of the dual beam anemometer is given in the figure above.

First reported by Goldstein and Hagen in 1967. The important features of the dual beam anemometer are the following:

1. Beam splitter splits laser output into two beams.
2. The two beams are focused nominally onto the same region of flow.
3. One of the beams falls directly on the photo multiplier tube (PMT).
4. The other beam is partially scattered by the fluid.

5. Some of this scattered beam will travel in the same direction as the reference beam.

6. It will pass the aperture and mix with the reference beam.

Fringe Model of the LDA

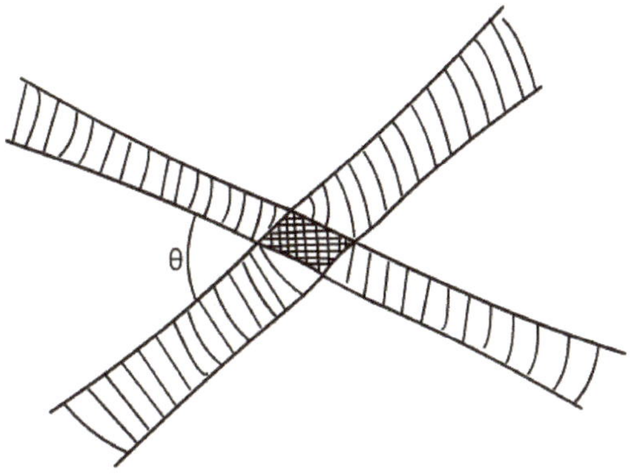

Fringe model of LDA

λ is the wavelength and θ is the angle between the beams. Where the two beams overlap, the two beams will interfere and produce a set of fringes. Distance between fringes $d = \frac{\lambda}{2} / \sin\frac{\theta}{2}$.

A particle crossing the section block the light. If one collects the light, it will fluctuate at the rate the particle crosses. This will modulate the light at a frequency of V/d where, V is the velocity of light particle.

$$f_D = \frac{v}{d} = \frac{2v}{\lambda}\sin\frac{\theta}{2} \qquad \frac{2v\sin\frac{\theta}{2}}{\lambda}$$

Relativistic Doppler Effect

Diagram 1. A source of light waves moving to the right, relative to observers, with velocity 0.7c. The frequency is higher for observers on the right, and lower for observers on the left.

The relativistic Doppler effect is the change in frequency (and wavelength) of light, caused by the relative motion of the source and the observer (as in the classical Doppler effect), when taking into account effects described by the special theory of relativity.

The relativistic Doppler effect is different from the non-relativistic Doppler effect as the equations include the time dilation effect of special relativity and do not involve the medium of propagation as a reference point. They describe the total difference in observed frequencies and possess the required Lorentz symmetry.

Visualization

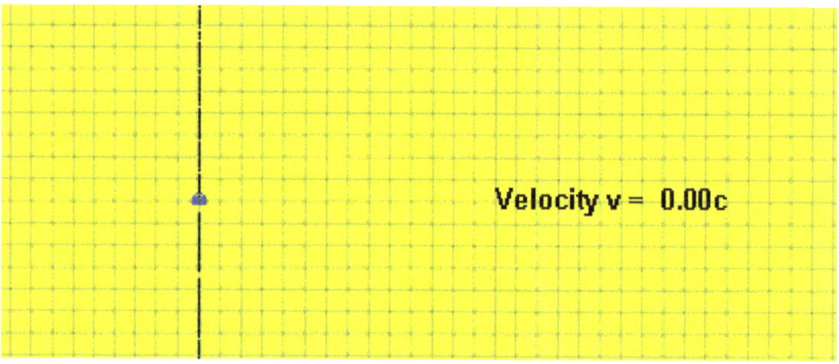

Demonstration of aberration of light and relativistic Doppler effect.

In Diagram, the blue point represents the observer, and the arrow represents the observer's velocity vector relative to its surroundings. When the observer is stationary, the x,y-grid appears yellow and the y-axis appears as a black vertical line. Increasing the observer's velocity to the right shifts the colors and the aberration of light distorts the grid. When the observer looks forward (right on the grid), points appear green, blue, and violet (blueshift) and grid lines appear farther apart. If the observer looks backward (left on the grid), then points appear red (redshift) and lines appear closer together. The grid has not changed, but its *appearance* for the observer has.

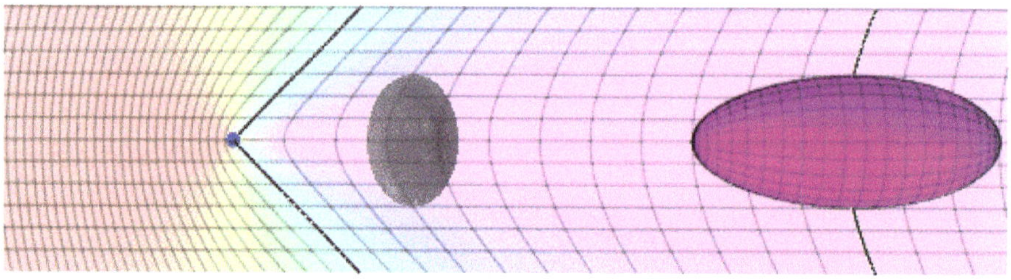

Diagram: The grey ellipse is a sphere moving relativistically at a constant velocity relative to an observer (blue dot); its oblate shape, as seen from our perspective, is due to Lorentz contraction. The colored ellipse is the sphere as seen by the observer. The background curves represent a grid (in xy coordinates) that is rigidly linked to the sphere; it is shown only at one moment in time.

Diagram illustrates that the grid distortion is a relativistic optical effect, separate from the underlying Lorentz contraction which is the same for an object moving toward an observer or away.

Analogy

Understanding relativistic Doppler effect requires understanding the Doppler effect, time dilation, and the aberration of light. As a simple analogy of the Doppler effect, consider two people playing catch. Imagine that a stationary pitcher tosses one ball each second (1 Hz) at one meter per second to a catcher who is standing still. The stationary catcher will receive one ball per second (1 Hz). Then the catcher walks away from the pitcher at 0.5 meters per second and catches a ball every 2 seconds (0.5 Hz). Finally, the catcher walks towards the pitcher at 0.5 meters per second and catches three balls every two seconds (1.5 Hz). The same would be true if the pitcher moved toward or away from the catcher. By analogy, the relativistic Doppler effect shifts the frequency of light as the emitter or observer moves toward or away from the other.

To understand the aberration effect, again imagine two people playing catch on two parallel conveyor belts (moving sidewalks) moving in opposite direction. The pitcher must aim differently depending on the speed and the spacing of the belts, and where the catcher is. The catcher will see the balls coming at a different angle than the pitcher chose to throw them. These angle changes depend on: 1) the instantaneous angle between the pitcher-catcher line and the relative velocity vector, and 2) the pitcher-catcher velocity relative to the speed of the ball. By analogy, the aberration of light depends on: 1) the instantaneous angle between the emitter-observer line and the relative velocity vector, and 2) the emitter-observer velocity relative to the speed of light.

Motion Along the Line of Sight

Assume the observer and the source are moving *away* from each other with a relative velocity v (v is *negative* if the observer and the source are moving *towards* each other). Considering the problem in the reference frame of the source, suppose one wavefront arrives at the observer. The next wavefront is then at a distance $\lambda = c / f_s$ away from the observer (where λ is the wavelength, f_s is the frequency of the wave the source emitted, and c is the speed of light).

The wavefront moves with velocity c, but at the same time the observer moves away with velocity v, so $\lambda + vt = ct$. This gives us

$$t = \frac{\lambda}{c-v} = \frac{c}{(c-v)f_s} = \frac{1}{(1-\beta)f_s},$$

where $\beta = v / c$ is the velocity of the observer in terms of the speed of light.

Due to the relativistic time dilation, the observer will measure this time to be

$$t_o = \frac{t}{\gamma},$$

where

$$\gamma = \frac{1}{\sqrt{1-\beta^2}}$$

is the Lorentz factor. The corresponding observed frequency is

$$f_o = \frac{1}{t_o} = \gamma(1-\beta)f_s = \sqrt{\frac{1-\beta}{1+\beta}}f_s.$$

The ratio

$$\frac{f_s}{f_o} = \sqrt{\frac{1+\beta}{1-\beta}}$$

is called the Doppler factor of the source relative to the observer.

The corresponding wavelengths are related by

$$\frac{\lambda_o}{\lambda_s} = \frac{f_s}{f_o} = \sqrt{\frac{1+\beta}{1-\beta}},$$

and the resulting redshift

$$z = \frac{\lambda_o - \lambda_s}{\lambda_s} = \frac{f_s - f_o}{f_o}$$

can be written as

$$z = \sqrt{\frac{1+\beta}{1-\beta}} - 1.$$

In the non-relativistic limit (when $v \ll c$) this redshift can be approximated by

$$z \ll \beta = \frac{v}{c},$$

corresponding to the classical Doppler effect.

Systematic Derivation for Inertial Observers

Let us repeat the derivation more systematically in order to show how the Lorentz equations can be used explicitly to derive a relativistic Doppler shift equation for waves that themselves are not relativistic.

Let there be two inertial frames of reference, S and S', constructed so that the axes of S and S' coincide at $t = t' = 0$, where t is the time as measured in S and t' is the time as measured in S'. Let S' be in motion relative to S with constant velocity v; without loss of generality, we will take this motion to be directed only along the x-axis. Thus, the Lorentz transformation equations take the form

$$x = \gamma\left(x' + \beta ct'\right) y = y'$$

$$z = z'$$

$$ct = \gamma\left(ct' + \beta x'\right)$$

$$\frac{dx}{dt} = \frac{v + \frac{dx'}{dt'}}{1 + \frac{v}{c^2}\frac{dx'}{dt'}}.$$

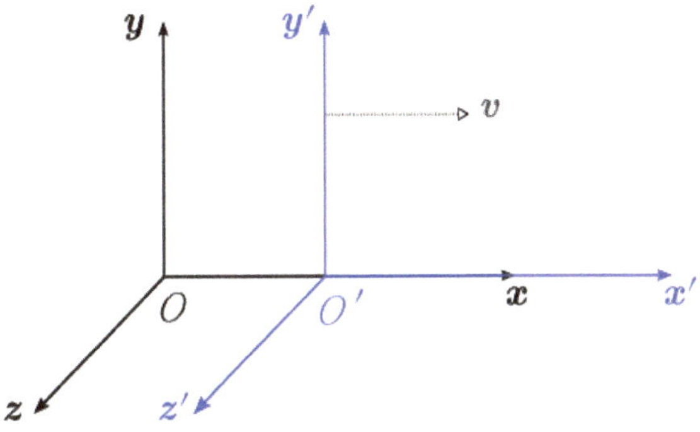

The system set up here is the standard one used in most physics text books, and the derivation presented here uses this standard construction. The S' reference frame moves with speed v relative to the S reference frame along the positive x-axis, and the reference frames coincide completely at $t = t' = 0$. Note that the signal emitter in S' is receding from the observer at O for positive v, by construction. The case where the signal emitter approaches the observer at O can be obtained by taking $v \to -v$ at the end of the derivation.

Velocity-addition formula, where $\beta = v/c$ and $\gamma = (1-\beta^2)^{-\frac{1}{2}}$, and c is the speed of light in a vacuum.

The derivation begins with what the observer in S' trivially sees. We imagine a signal source is positioned stationary at the origin, O', of the S' system. We will take this signal source to produce its first pulse at time $t_1' = 0$ (this is *event 1*) and its second pulse at time $t_2' = 1/f'$ (this is *event 2*), where f' is the frequency of the signal source as the observer in S' reckons it. We then simply use the Lorentz transformation equations to see when and where the observer in S sees these two events as occurring:

	Observer in S'	**Observer in S**
Event 1	$x_1' = 0$ $t_1' = 0$	$x_1 = 0$ $t_1 = 0$
Event 2	$x_2' = 0$ $t_2' = \frac{1}{f'}$	$x_2 = \gamma\frac{v}{f'}$ $t_2 = \gamma\frac{1}{f'}$

The period between the pulses as measured by the S observer is *not*, however, $t_2 - t_1$ because *event 2* occurs at a different point in space to *event 1* as observed by the S observer (that is, $x_2 \neq x_1$) — we must factor in the time taken for the pulse to travel from x_2 to x_1. Note that this complication is *not* relativistic in nature: this is the ultimate cause of the Doppler effect and is also present in the classical treatment. This transit time is equal to the difference $x_2 - x_1$ divided by the speed of the pulse as the S observer sees it. If the pulse moves at speed $-u'$ in S' (negative because it moves in the negative x-direction, towards the S observer at O), then the speed of the pulse moving *towards* the observer at O, as S sees it, is:

$$-u = \frac{-u' + v}{1 + (-u')\dfrac{v}{c^2}},$$

using the Lorentz equation for the velocities, above. Thus, the period between the pulses that the observer in S measures is:

$$\tau = t_2 - t_1 + \gamma \frac{v}{f'} \left(\frac{u' - v}{1 - \dfrac{vu'}{c^2}} \right)^{-1}$$

$$= \frac{\gamma}{f'} + \frac{\gamma}{f'} \frac{v}{u' - v} \left(1 - \frac{vu'}{c^2} \right).$$

Replacing τ with $1/$ and simplifying, we get the required result that gives the relativistic Doppler shift of *any* moving wave in terms of the stationary frequency, f':

$$f = \gamma \left(1 - \frac{v}{u'} \right) f'.$$

Ignoring the relativistic effects by taking v ≪ c or $c \to \infty$ (equivalent to $\gamma \to 1$) gives the classical Doppler formula:

$$f = \left(1 - \frac{v}{u'} \right) f'.$$

For electromagnetic radiation where $u' = c$ the formula becomes

$$f = \gamma \left(1 - \frac{v}{c} \right) f' = \gamma (1 - \beta) f' = f' \sqrt{\frac{1 - \beta}{1 + \beta}}$$

or in terms of wavelength:

$$\lambda = \lambda' \sqrt{\frac{1 + \beta}{1 - \beta}},$$

where λ' is the wavelength of the source at the origin O' as the observer in S' sees it. In these equations v (and thus β) is assumed positive when the source is receding from the observer, and negative when approaching.

For electromagnetic radiation, the limit to classical mechanics, $c \to \infty$, is instructive. The Doppler effect formula simply becomes $f = f'$. This is the "correct" result for classical mechanics, although it is clearly in disagreement with experiment. It is "correct" since classical mechanics regards the maximum speed of interaction — for electrodynamics, the speed of light — to be infinite. The Doppler effect, classical or relativistic, occurs because the wave source has time to move by the time that previous waves encounter the observer. This means that the subsequent waves are emitted further away (or closer) to the observer than they otherwise would be if the source were not in motion. The effect of this is to stretch (or compress) the wavelength of the wave as the observer encounters them. If however the waves travel instantaneously, the fact that the source is further away (or closer) makes no difference because the waves arrive at the observer no later or earlier than they would anyway since they arrive instantaneously. Thus, classical mechanics predicts that there should be no Doppler effect for light waves, whereas the relativistic theory gives the correct answer, as confirmed by experiment.

Transverse Doppler Effect

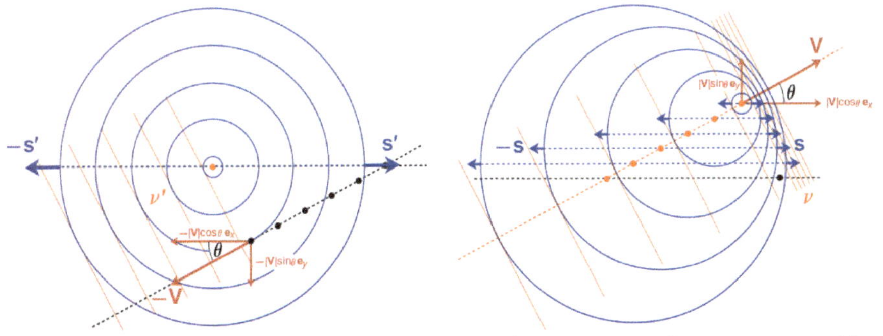

The relativistic Doppler effect, in two spatial dimensions, where the observer and emitter move with relative velocity V as shown, and the circles are wave fronts each a cycle apart. The medium the waves propagate through is stationary relative to the emitter. Left: In the emitter's frame, the wave has frequency ν' propagates with wave velocity s′, measured in the direction shown with the magnitude equal to speed of propagation. The observer will appear to move with velocity −V. Right: In the observer's frame, the emitter will appear to move with velocity V, the frequency ν increases during approach to the observer, and the wave velocity s is given by the relativistic velocity addition formula.

The *transverse Doppler effect* is the nominal redshift or blueshift predicted by special relativity that occurs when the emitter and receiver are at the point of closest approach. Light emitted at closest approach in the source frame will be blueshifted at the receiver. Light received at closest approach in the receiver frame will be redshifted relative to its source frequency.

Assuming the objects are not accelerated, light emitted when the objects are closest together will be received some time later. At reception, the amount of redshift will be

$$\frac{1}{\gamma} = \sqrt{1 - \frac{v^2}{c^2}}.$$

If the light is received when the objects are closest together, then it was emitted some time earlier. At reception, the amount of blueshift will be

$$\gamma = \frac{1}{\sqrt{1 - \frac{v^2}{c^2}}}.$$

Classical theory does not make a specific prediction for either of these two cases, as the shift depends on the motions relative to the medium.

The transverse Doppler effect is a consequence of the relativistic Doppler effect.

$$f_o = \frac{f_s}{\gamma\left(1+\dfrac{v}{c}\cos\theta_o\right)}$$

In the frame of the receiver, θ_0 represents the angle between the direction of the emitter at emission, and the observed direction of the light at reception. In the case when $\theta_0 = \pi/2$, the light was emitted at the moment of closest approach, and one obtains the transverse redshift

$$f_o = \frac{f_s}{\gamma}$$

The transverse Doppler effect is one of the main novel predictions of the special theory. As Einstein put it in 1907: according to special relativity the moving object's emitted frequency is reduced by the Lorentz factor, so that the received frequency is reduced by the same factor.

Reciprocity

Sometimes the question arises as to how the transverse Doppler effect can lead to a redshift as seen by the "observer" whilst another observer moving with the emitter would *also* see a redshift of light sent (perhaps accidentally) from the receiver.

It is essential to understand that the concept "transverse" is not reciprocal. Each participant understands that when the light reaches them transversely as measured in terms of that person's rest frame, the other had emitted the light afterward as measured in the other person's rest frame. In addition, each participant measures the other's frequency as reduced ("time dilation"). These effects combined make the observations fully reciprocal, thus obeying the principle of relativity.

Experimental Verification

In practice, experimental verification of the transverse effect usually involves looking at the longitudinal changes in frequency or wavelength due to motion for approach and recession: by comparing these two ratios together we can rule out the relationships of "classical theory" and prove that the real relationships are "redder" than those predictions. The transverse Doppler shift is central to the interpretation of the peculiar astrophysical object SS 433.

The first longitudinal experiments were carried out by Herbert E. Ives and Stilwell in (1938), and many other longitudinal tests have been performed since with much higher precision. Also a direct transverse experiment has verified the redshift effect for a detector actually *aimed* at 90 degrees to the object.

Motion in an Arbitrary Direction

If, in the reference frame of the observer, the source is moving away with velocity v at an angle θ_o relative to the direction from the observer to the source (at the time when the light is emitted), the frequency changes as

$$f_o = \frac{f_s}{\gamma\left(1+\frac{v}{c}\cos\theta_o\right)}. \tag{14}$$

In the particular case when $\theta_o = 90^\circ$ and $\cos\theta_o = 0$ one obtains the transverse Doppler effect:

$$f_o = \frac{f_s}{\gamma}.$$

Due to the finite speed of light, the light ray (or photon, if you like) perceived by the observer as coming at angle θ_o, was, in the reference frame of the source, emitted at a different angle θ_s. $\cos\theta_o$ and $\cos\theta_s$ are tied to each other via the relativistic aberration formula:

$$\cos\theta_o = \frac{\cos\theta_s - \frac{v}{c}}{1-\frac{v}{c}\cos\theta_s}.$$

Therefore, Eq. (14) can be rewritten as

$$f_o = \gamma\left(1-\frac{v\cos\theta_s}{c}\right)f_s. \tag{15}$$

For example, a photon emitted at the right angle in the reference frame of the emitter ($\cos\theta_s = 0$) would be seen blue-shifted by the observer:

$$f_o = \gamma f_s.$$

In the non-relativistic limit, both formulæ (14) and (15) give

$$\frac{\Delta f}{f} \simeq -\frac{v\cos\theta}{c}.$$

Doppler Effect on Intensity

The Doppler effect (with arbitrary direction) also modifies the perceived source intensity: this can be expressed concisely by the fact that source strength divided by the cube of the frequency is a Lorentz invariant (here, "source strength" refers to spectral intensity in *frequency*, i.e., power per unit solid angle and per unit frequency, expressed in watts per steradian per hertz; for spectral intensity in *wavelength*, the cube should be replaced by a fifth power). This implies that the total radiant intensity (summing over all frequencies) is multiplied by the fourth power of the Doppler factor for frequency.

As a consequence, since Planck's law describes the black body radiation as having a spectral intensity in frequency proportional to $\nu^3 / \left(e^{h\nu/kT} - 1\right)$ (where T is the source temperature and ν the frequency), we can draw the conclusion that *a black body spectrum seen through a Doppler shift (with arbitrary direction) is still a black body spectrum* with a temperature multiplied by the same Doppler factor as frequency.

Accelerated Motion

For general accelerated motion, or when the motions of the source and receiver are analyzed in an arbitrary inertial frame, the distinction between source and receiver motion must again be taken into account.

The Doppler shift when observed from an arbitrary inertial frame:

$$\frac{f_o}{f_s} = \frac{1 - \frac{\|\vec{v_o}\|}{\|\vec{c}\|} cos(\theta_{co})}{1 - \frac{\|\vec{v_s}\|}{\|\vec{c}\|} cos(\theta_{cs})} \sqrt{\frac{1-\left(\frac{v_s}{c}\right)^2}{1-\left(\frac{v_o}{c}\right)^2}}$$

where:

$\vec{v_s}$ is the velocity of the source at the time of emission

$\vec{v_o}$ is the velocity of the receiver at the time of reception

$\vec{c}$ is the light velocity vector

θ_{cs} is the angle between the source velocity and the light velocity at the time of emission

θ_{co} is the angle between the receiver velocity and the light velocity at the time of reception

If $\vec{c}$ is parallel to $\vec{v_s}$, then $\theta_{cs} = 0^\circ$, which causes the frequency measured by the receiver f_o to increase relative to the frequency emitted at the source f_s. Similarly, if $\vec{c}$ is anti-parallel to $\vec{v_s}$, $\theta_{cs} = 180^\circ$, , which causes the frequency measured by the receiver f_o to decrease relative to the frequency emitted at the source f_s.

This is the classical Doppler effect multiplied by the ratio of the receiver and source Lorentz factors.

Due to the possibility of refraction, the light's direction at emission is generally not the same as its direction at reception. In refractive media, the light's path generally deviates from the straight distance between the points of emission and reception. The Doppler effect depends on the component of the emitter's velocity parallel to the light's direction at emission, and the component of the receiver's velocity parallel to the light's direction at absorption. This does not contradict Special Relativity.

The transverse Doppler effect can be analyzed from a reference frame where the source and receiver have equal and opposite velocities. In such a frame the ratio of the Lorentz factors is always 1, and all Doppler shifts appear to be classical in origin. In general, the observed frequency shift is an invariant, but the relative contributions of time dilation and the Doppler effect are frame dependent.

References

- Durst, F; Melling, A. and Whitelaw, J. H. (1976) Principles and Practice of Laser Doppler Anemometry, Academic Press, London, ISBN 0-12-225250-0
- Chao, Mayer (1971). "An Additional Effect of Tropospheric Refraction on the Radio Tracking of Near-Earth Spacecraft at Low Elevation Angles" (PDF). Retrieved 6 August 2015
- Goode, RL; Ball, G; Nishihara, S; Nakamura, K (1996). "Laser Doppler vibrometer (LDV)--a new clinical tool for the otologist". The American journal of otology. 17 (6): 813–22. PMID 8915406
- Feynman, Richard P.; Leighton, Robert B.; Sands, Matthew (February 1977). The Feynman Lectures on Physics: Volume 1. Reading, Massachusetts: Addison-Wesley. pp. 34–7 f. ISBN 9780201021165. LCCN 2010938208
- Ives, H. E.; Stilwell, G. R. (1938). "An experimental study of the rate of a moving atomic clock". Journal of the Optical Society of America. 28 (7): 215. Bibcode:1938JOSA...28..215I. doi:10.1364/JOSA.28.000215
- Landau, L.D.; Lifshitz, E.M. (2005). The Classical Theory of Fields. Course of Theoretical Physics: Volume 2. Trans. Morton Hamermesh (Fourth revised English ed.). Elsevier Butterworth-Heinemann. pp. 1–3. ISBN 9780750627689
- "ALHAT Detects Landing Hazards on the Surface". Research News, Langley Research Center. NASA. Retrieved February 8, 2013